Status of Recent Geoscience Graduates 2017

Carolyn Wilson

American Geosciences Institute

Alexandria, VA 22302

Status of Recent Geoscience Graduates 2017
Edited by Carolyn Wilson
ISBN: 978-0-922152-64-3

Graphs by Carolyn Wilson, AGI Workforce Program
Design by Brenna Tobler, AGI Graphic Designer

For more information on the American Geosciences Institute and its publications check us out at store.americangeosciences.org.

Carolyn Wilson, Geoscience Workforce Data Analyst
American Geosciences Institute
4220 King Street, Alexandria, VA 22302

www.americangeosciences.org
cwilson@americangeosciences.org
(703) 379-2480, ext. 632

Front cover photo: A student taking depth and velocity of flow at an exposed pipe using a Flow Mate during an internship with Orange County Environmental Resources in the Newport Bay Watershed. Credit: Mayra Martinez for AGI's 2017 Life as a Geoscientist contest

All photos in this report were submitted to the 2017 Life as a Geoscientist contest, which requested images representing meaningful geoscience work through internships, research, employment, or field experiences.

american geosciences institute
connecting earth, science, and people

GEOSCIENCE STUDENT EXIT SURVEY

About AGI's Geoscience Student Exit Survey

The American Geosciences Institute (AGI) launched the Geoscience Student Exit Survey to help geoscience departments to assess the educational experiences of graduating students, as well as for AGI to understand the trends of strengths and weaknesses of new graduates entering the workforce. With this survey, we can identify student decision points for entering and persisting in a geoscience field, measure participation in co-curricular activities that support the development of critical geoscience skills, identify the geoscience fields of interest, identify the preferred jobs and industries of graduating students, and establish a benchmark for a detailed study of career pathways of early career geoscientists.

This report examines the responses to AGI's Geoscience Student Exit Survey by graduates from the 2016-2017 academic year. This was the fifth year of data collection for this project.

The survey has four major sections: student demographics, educational background, postsecondary education experiences, and post-graduation plans, with specific questions covering community college experiences, quantitative skills, field and research experiences, internships, and details about their immediate plans for graduate school or a new job. The survey was piloted twice in Spring 2011 and Spring 2012. For Spring 2013 graduation, we opened the survey up to all geoscience departments in the United States. Since 2014, the survey was available for graduates at the end of each semester—fall, spring, and summer. AGI also asks its member societies to send the survey out to their student membership, which, in addition to the numerous departments already distributing the survey, reaches a larger pool of recent geoscience graduates.

As awareness of the survey grows, AGI has been able to engage collaborations with Canada and the United Kingdom to send versions of the survey to their graduates as well. AGI will continue to work to expand the number of countries that distribute the survey to gain a more global understanding of the preparation of geoscience graduates for the workforce.

To encourage participation, departments, member societies, and international organizations that distribute AGI's Geoscience Student Exit Survey will receive the data in aggregate for their constituency, as long as they have a sufficient number of participating students to ensure individual response privacy.

If you would like more information or would like your department, society, or country to participate in AGI's Geoscience Student Exit Survey, please contact Carolyn Wilson at cwilson@americangeosciences.org.

Acknowledgements

I would like to recognize a few organizations and individuals for their support for this project. Thanks to ConocoPhillips for their monetary contribution towards the project this year. Thanks also to the American Geophysical Union, the American Institute of Professional Geologists, the Geological Society of America, and the Society of Exploration Geophysicists for distributing the survey to their student membership. I also want to thank the AGI Workforce 2017 Fall Intern, Caroline Kelleher, for her hard work cleaning and organizing the data from this survey, as well as her help creating the map figures in the report. Finally, I would especially like to thank the department contacts from each participating department for distributing the survey to their graduating students.

Executive Summary

The American Geosciences Institute's (AGI) Status of Recent Geoscience Graduates 2017 provides an overview of the demographics, activities, and experiences of geoscience degree recipients during the 2016-2017 academic year. This research draws attention to student preparation in the geosciences, their education and career path decisions, as well as examines some of the questions raised about student transitions into the workforce. This is the fifth consecutive year of this survey and report, and with this release, we are starting to detail emerging trends in the experiences of postsecondary geoscience students.

Major findings from the Status of Recent Geoscience Graduates 2017 report include:

- The demographics of geoscience degree recipients has remained steady, with female participation of at least 40 percent at all degree levels, but underrepresented minorities participation remains at or below 12%. However, percentages of graduates that are unwilling to provide their demographic information has been steadily increasing since 2014.

- Over the past five years, the majority of geoscience graduates complete Calculus II as their highest quantitative course, but participation in higher quantitative courses such as differential equations and linear algebra is low, with as few as 25 percent of bachelor's recipients having taken those courses, and graduate degree participation levels are marginally higher. This quantitative skill deficiency is viewed as a critical negative impact on employment resilience for many new graduates.

- Nearly every graduate participated in at least one field or research experience before graduation, and most students participate in multiple field and research experiences.

- Internships provide graduates with critical professional development skills and experiences, but only 40% of bachelor's graduates, 68% of master's graduates, and 48% of doctoral graduates participated in an internship before graduation, consistent with the trend over the past five years.

- Hiring of recent graduates at graduation is at its lowest in 5 years for all degree levels. This is particularly pronounced for doctoral graduates dropping from 70% in 2014 to 36% in 2017. A number of factors are likely at play, including a slow recovery in the resources industry, displacement of some jobs with automation, and uncertainty in the regulatory environment with the current Administration.

- 2017 had the lowest percentage in the last five years of bachelor's graduates (35%) and the highest percentage of master's graduates (31%) planning to attend graduate school immediately after graduation. Even though graduate programs have generally been at full capacity, recent slowing in hiring of new graduates is encouraging recent graduates to consider another degree delaying entry into the workforce.

Geoscience graduates at all degree levels tend to receive a strong technical background in geoscience. But after a number of years of strong hiring, there is a marked slow-down in hiring of new graduates. There may be a number of factors leading to this slow down, including softness in hiring in the resources industry, uncertainty about environmental regulations, and the emergence of automation offsetting some middle-skill geoscience jobs. Based on surveys of prior graduates, students with strong quantitative skills have fared well at gaining and retaining employment. Likewise, non-technical skills such as effective communications, as well as business and finance exposure, have been identified as key employer desires. However, these skills are not necessarily readily available within the formal degree programs. AGI's Geoscience Student Exit Survey is one of the few tools that help us track these changes as experienced by recent graduates, and AGI recognizes the importance of continuing this research study annually in order to provide this information to all within the geoscience community.

Contents

An Overview of the Demographics of the Participants

This year, AGI's Geoscience Student Exit Survey was made available to geoscience graduates at all traditional graduation periods (winter, spring, and, summer) during the 2016-2017 academic year, to be collectively referenced as "2017". Each spring, department heads and chairs are requested for their department's participation in the survey, and a running list of participating departments is updated each year. Each department is periodically reminded to share the survey with their graduating students near the end of each semester. As incentive to participate, AGI gives the departments the data in aggregate for their graduates for their own assessment purposes.

AGI has also enlisted the help of societies within AGI's Federation to send the survey to their student membership in the spring. In 2016, AGI received assistance with the survey distribution from the American Geophysical Union (AGU), the American Institute of Professional Geologists (AIPG), the Geological Society of America (GSA), and the Society of Exploration Geophysicists (SEG). These societies helped to recruit approximately 12 percent of the recent graduates that participated in the survey for 2017.

The survey was available to the winter and summer graduates for two months, and the spring graduates had three months to complete the survey. At the close of the survey, 513 graduating students from 147 geoscience schools or departments provided responses—419 bachelor's graduates, 58 master's graduates, and 36 doctoral graduates. Five states, Maine, Nevada, South Carolina, Vermont, and Wyoming, were not represented in this sample of geoscience graduates. This was an increase in participation from last year, but the increase was only seen in the number of bachelor's graduates. Fewer master's and doctoral graduates took the survey in 2017 than the 2016 survey.

The first section of the survey covered student demographics to establish an understanding of who graduated in the geosciences. The data remained fairly consistent with previous years, but there have been slight shifts in the gender distribution from year to year. In 2017, there were 5 percent more female bachelor's graduates and 6 percent more male master's graduates. While the gender distribution of graduates at the three degree levels have varied over

recent years, neither group has dropped below 40 percent representation within the graduates of each degree. The survey allows for graduates to identify with other gender categories, and a few bachelor's graduates identified themselves as gender non-binary or gender non-conforming. As in previous years, students indicating their citizenship as U.S. Citizen or Permanent Resident were asked to indicate their race and ethnicity. The percentage of underrepresented minorities contains African Americans, Hispanic/Latinos, Native Americans/Alaskans, and Native Hawaiians/Pacific Islanders. However, it is important to note that is the percentage of underrepresented minorities is dominated by the Hispanic/Latino population of geoscience graduates. The percentages of underrepresented minority geoscience graduates continue to remain at 12 percent or less depending on the degree awarded. The age distribution of graduates in 2016 is fairly similar to the distributions in previous years, with clear ranges of ages for graduates at the bachelor's and master's degree levels. The age distribution of doctoral graduates was more variable likely due to the smaller sample of doctoral graduates in 2017. Geoscience graduates over the age of 35 tend to complete master's or bachelor's degrees, likely in an attempt to gain more skills or work towards a career shift. It is important to note that over the past five years, the percentage of graduates that have been unwilling to provide some or all of their demographic information has increased.

For the 2015 survey, recent graduates were asked to report the highest education level of their parents or guardians. Concerns have been raised that geoscience programs tend to attract students from middle and upper class families, possibly due to parental familiarity with the subject area or the high cost of the activities associated with the degree. AGI has decided to use the highest education level of parents as a proxy for inferring the socioeconomic status of geoscience graduates. In 2017, 70 percent of bachelor's graduates, 70 percent of master's graduates, and 72 percent of doctoral graduates had at least one parent with a post-secondary degree. Among the master's graduates, that is a 16 percent increase compared to 2016. This question also indicated that 9 percent of bachelor's graduates, 5 percent of master's graduates, and 9 percent of doctoral graduates were first-generation college students.

Distribution of participating graduating students and departments*

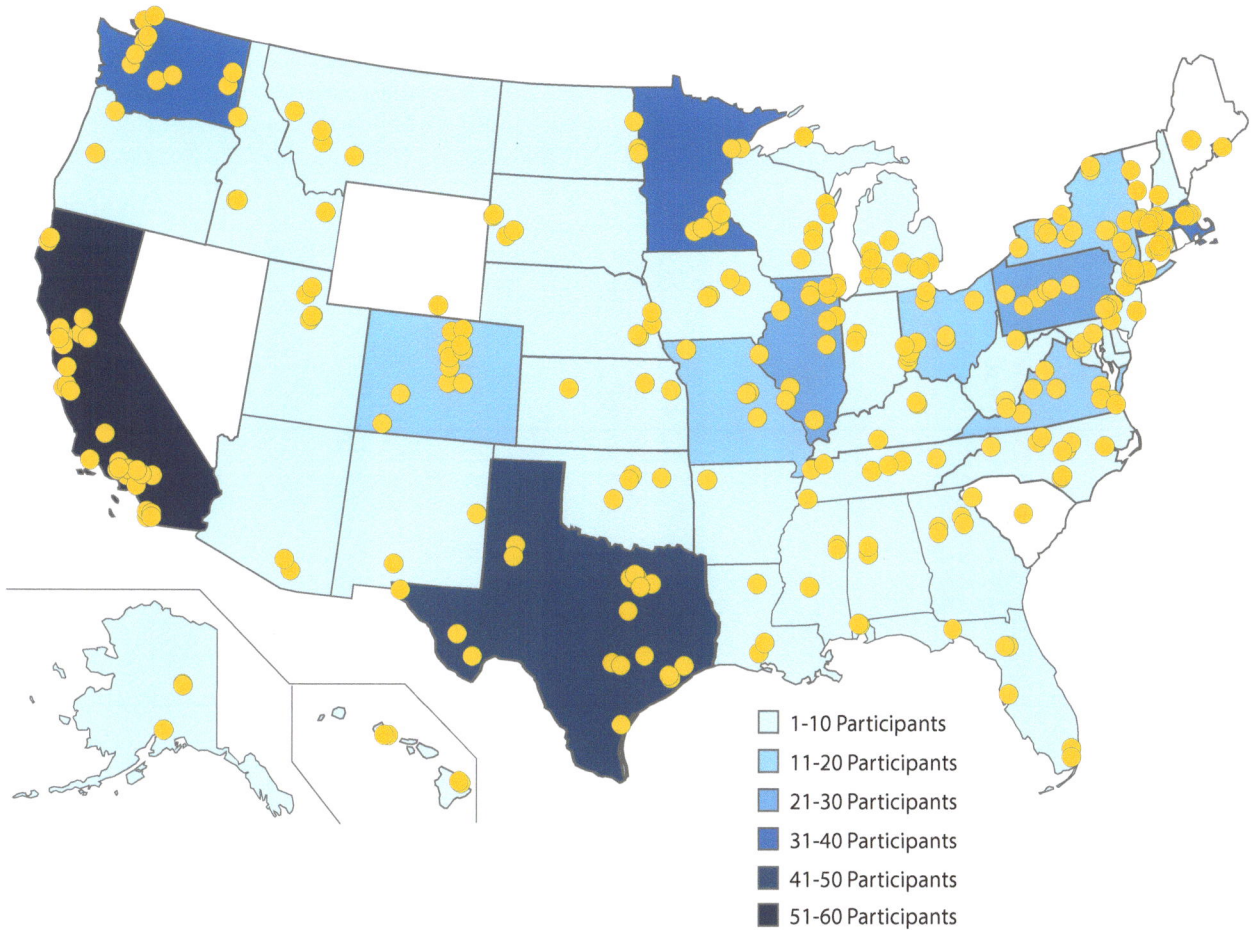

Legend:
- 1-10 Participants
- 11-20 Participants
- 21-30 Participants
- 31-40 Participants
- 41-50 Participants
- 51-60 Participants

The relative distribution by state of the universities and their graduating geoscience students across the United States that participated in the Exit Survey. *See Appendix I for list of departments

Degree received by participating graduates

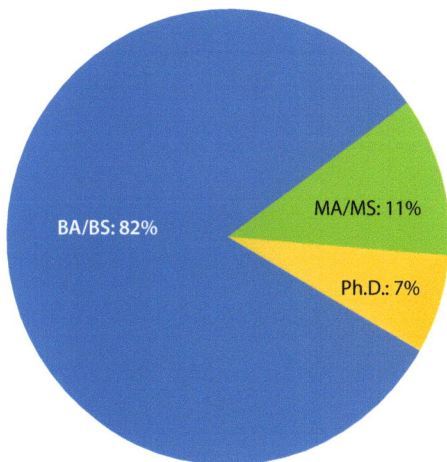

BA/BS: 82%
MA/MS: 11%
Ph.D.: 7%

Percentage of respondents within different classified institutions**

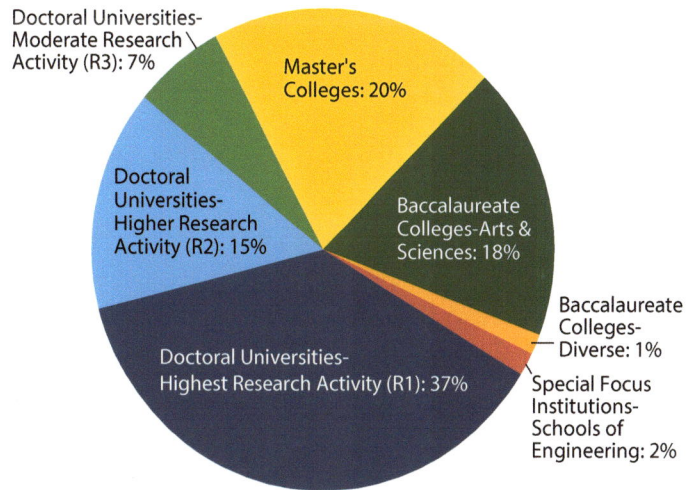

Doctoral Universities-Moderate Research Activity (R3): 7%
Master's Colleges: 20%
Doctoral Universities-Higher Research Activity (R2): 15%
Baccalaureate Colleges-Arts & Sciences: 18%
Baccalaureate Colleges-Diverse: 1%
Doctoral Universities-Highest Research Activity (R1): 37%
Special Focus Institutions-Schools of Engineering: 2%

**See Appendix II for definitions of the Carnegie University Classification System

Gender breakdown of graduates

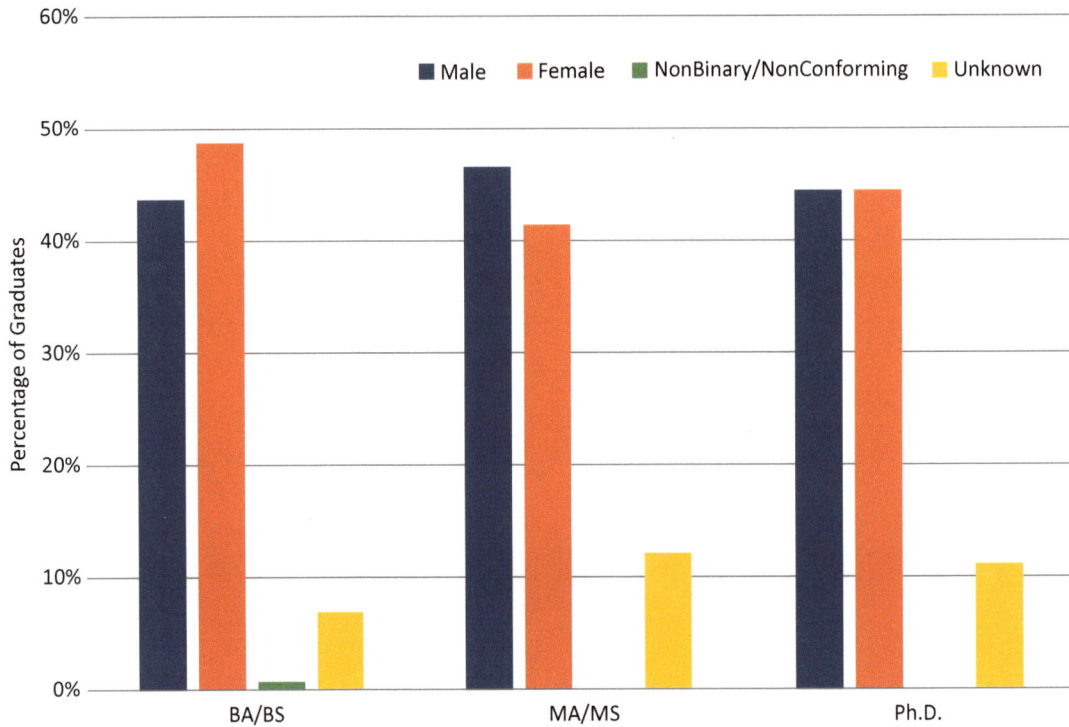

Age distribution of graduates

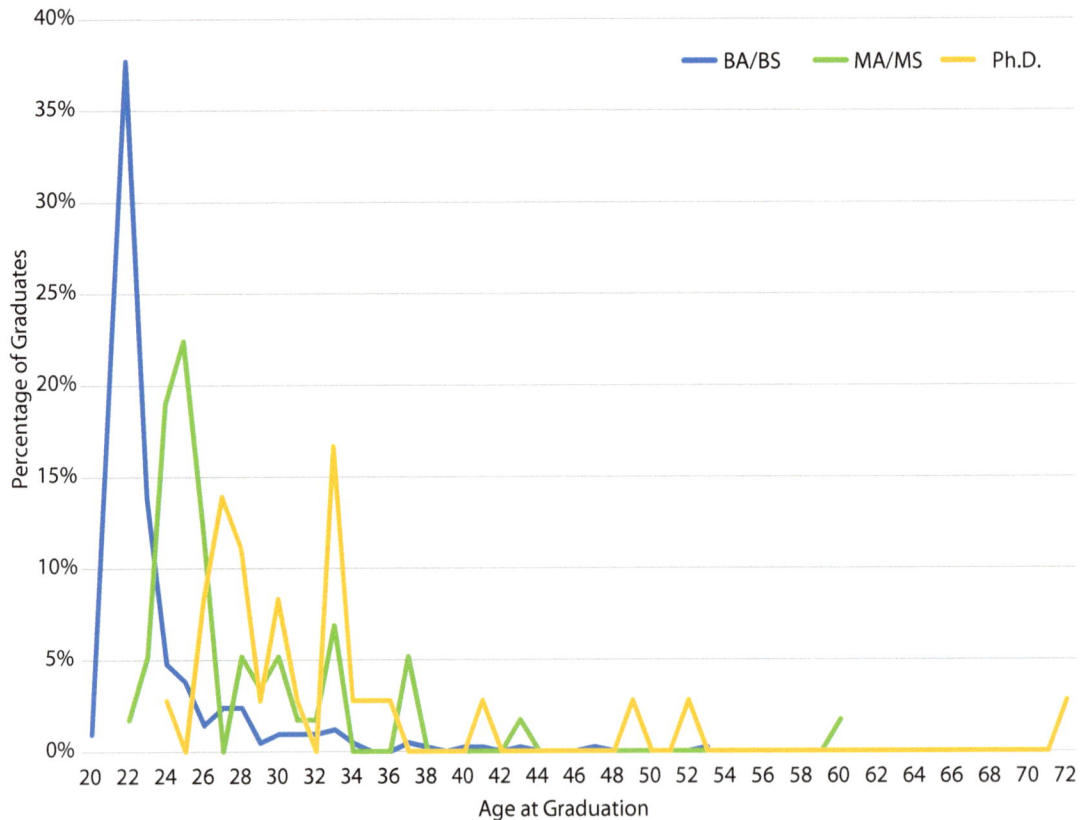

Citizenship of graduating students

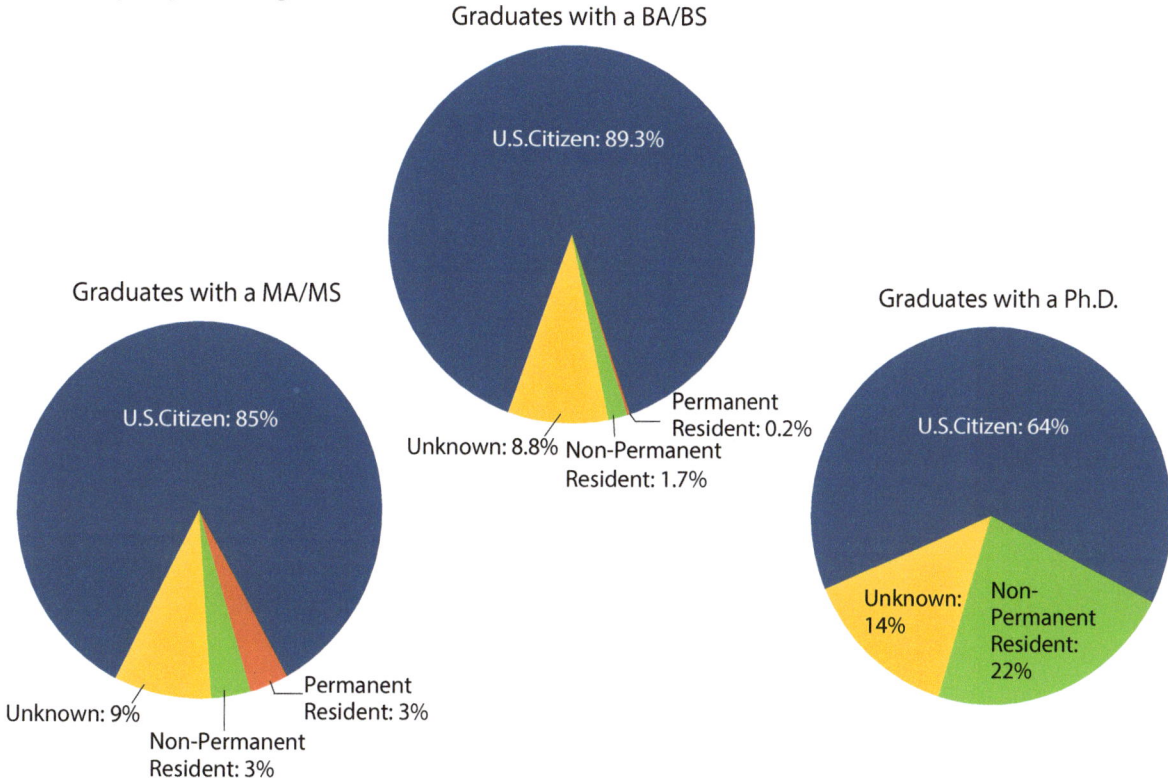

Graduates with a BA/BS

U.S.Citizen: 89.3%

Unknown: 8.8%

Permanent Resident: 0.2%

Non-Permanent Resident: 1.7%

Graduates with a MA/MS

U.S.Citizen: 85%

Unknown: 9%

Non-Permanent Resident: 3%

Permanent Resident: 3%

Graduates with a Ph.D.

U.S.Citizen: 64%

Unknown: 14%

Non-Permanent Resident: 22%

Race/ethnicity of graduating students

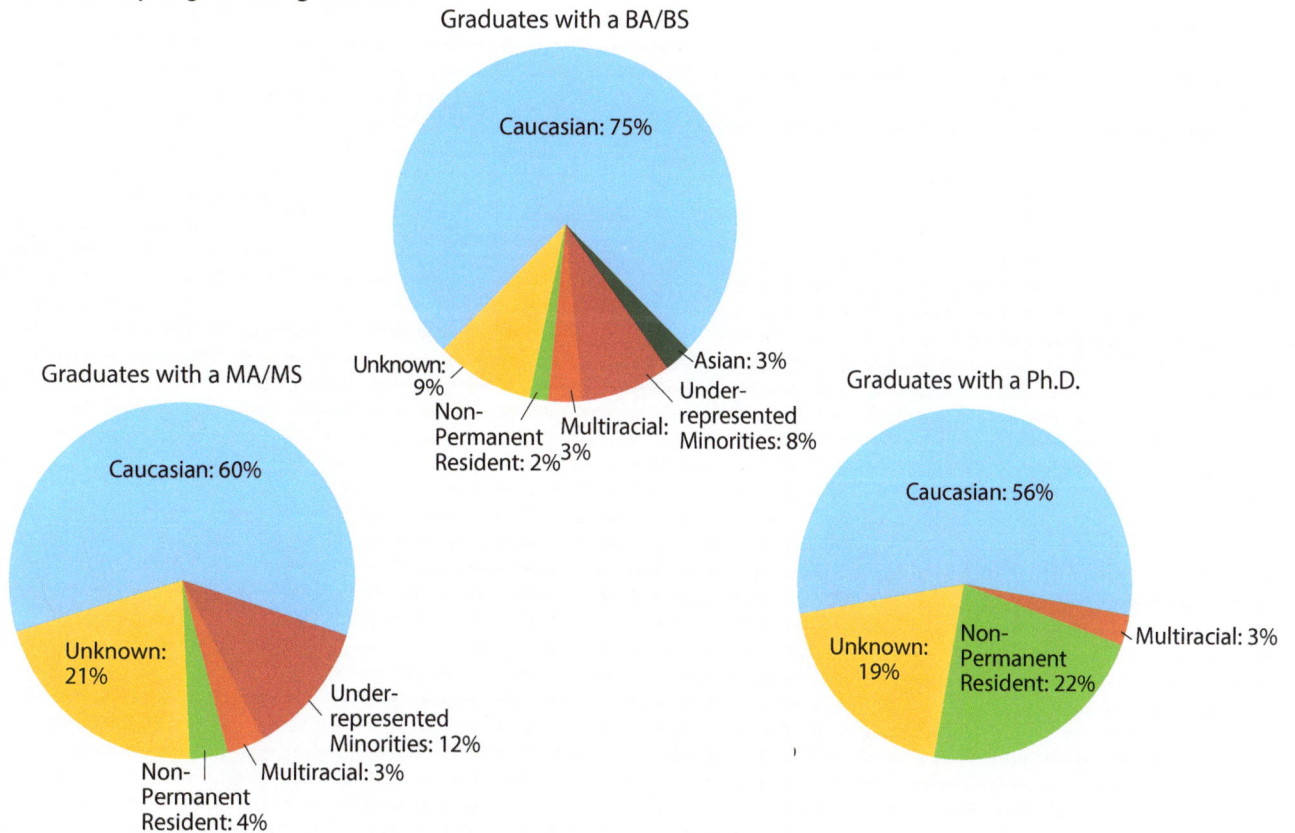

Graduates with a BA/BS

Caucasian: 75%

Unknown: 9%

Non-Permanent Resident: 2%

Multiracial: 3%

Under-represented Minorities: 8%

Asian: 3%

Graduates with a MA/MS

Caucasian: 60%

Unknown: 21%

Non-Permanent Resident: 4%

Multiracial: 3%

Under-represented Minorities: 12%

Graduates with a Ph.D.

Caucasian: 56%

Unknown: 19%

Non-Permanent Resident: 22%

Multiracial: 3%

Highest education level of a parent/guardian of graduates

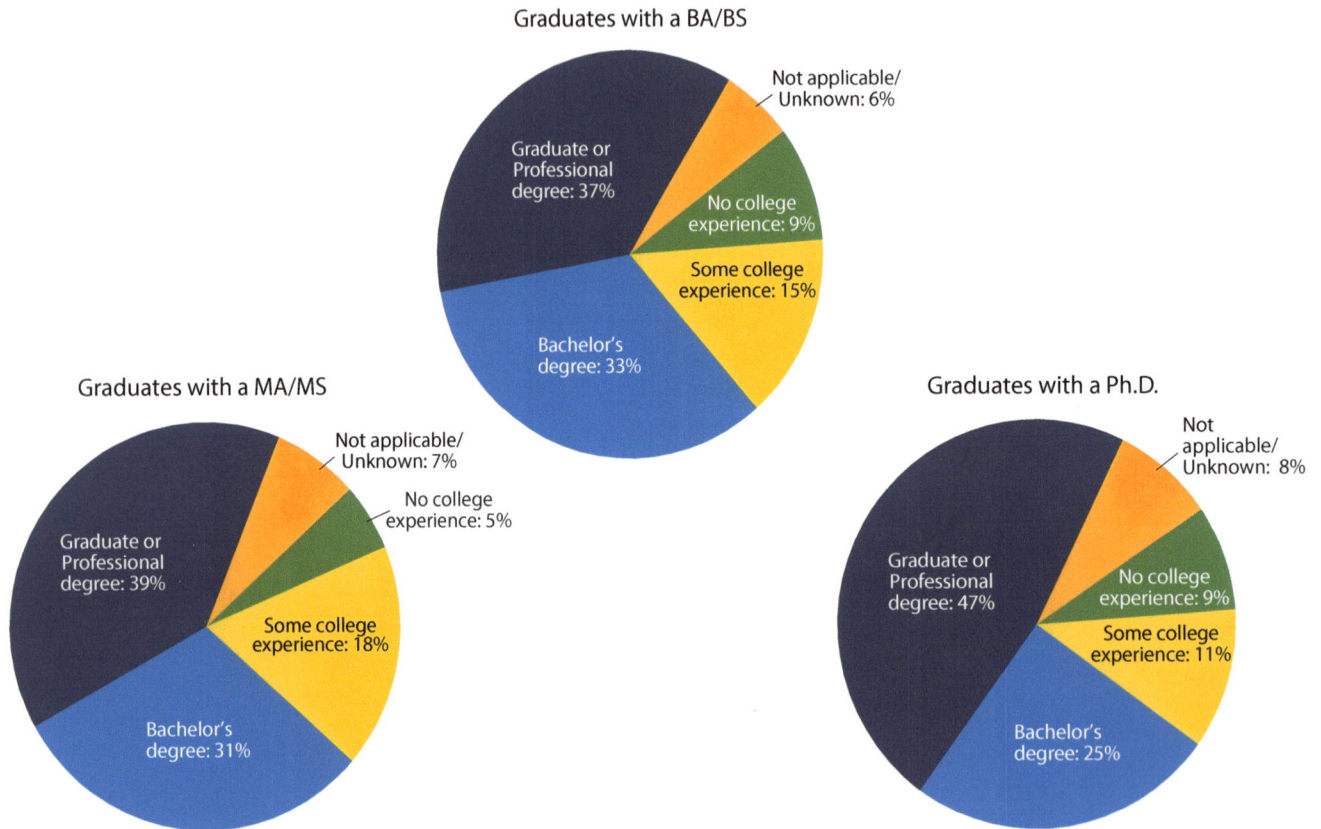

Graduates with a BA/BS

- Graduate or Professional degree: 37%
- Bachelor's degree: 33%
- Some college experience: 15%
- No college experience: 9%
- Not applicable/Unknown: 6%

Graduates with a MA/MS

- Graduate or Professional degree: 39%
- Bachelor's degree: 31%
- Some college experience: 18%
- No college experience: 5%
- Not applicable/Unknown: 7%

Graduates with a Ph.D.

- Graduate or Professional degree: 47%
- Bachelor's degree: 25%
- Some college experience: 11%
- No college experience: 9%
- Not applicable/Unknown: 8%

Christina Byrd for AGI's 2017 Life as a Geoscientist contest

At the Sternberg Museum of Natural History in Hays, KS, two college students digitize the invertebrate paleontology collection for future public viewing.

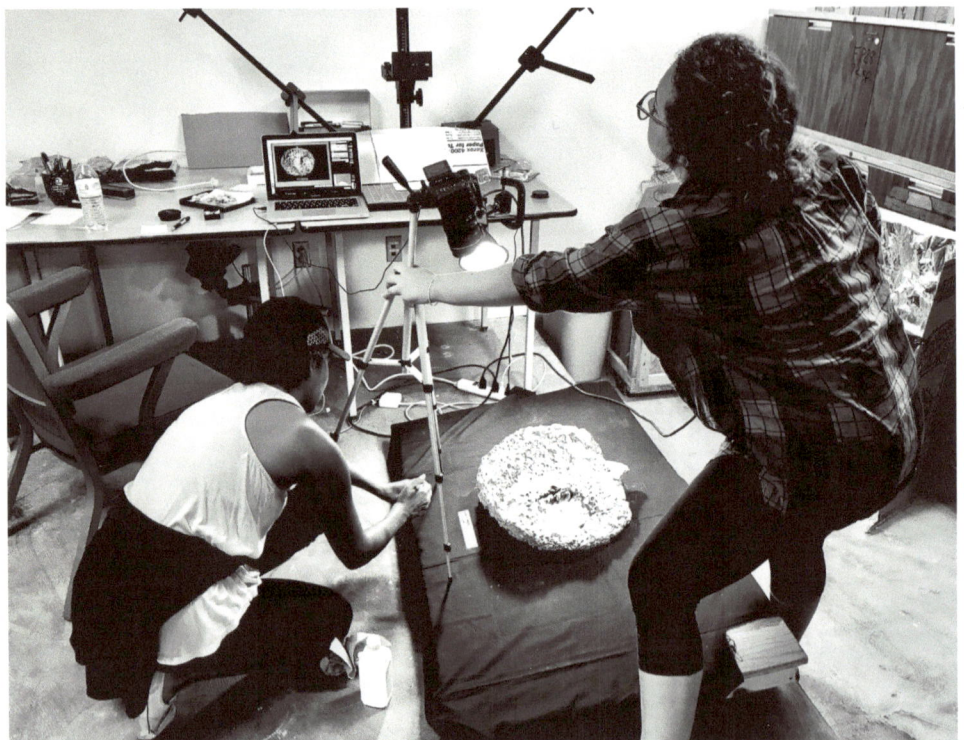

Nicholas Perez for AGI's 2017 Life as a Geoscientist contest
Clyde Findlay, Ph.D. student, and Keller Herrin, undergraduate research assistant, from Texas A&M University, conduct a drone survey near Castleton Tower, Utah. They will use the series of aerial images to create a 3D outcrop model, and combine the results with field measurements to investigate stratigraphic questions about the upper Paleozoic Cutler Group.

Quantitative Skills and Geoscience Background of the Graduating Students

This section examines graduates' educational background, such as quantitative rigor, the role of K-12 experiences, and the importance of two-year colleges.

The students were asked to select all of the quantitative courses they have taken at a two-year or four-year institution. Consistently over the past five years, the majority of geoscience graduates, regardless of degree, complete Calculus II as their highest quantitative course. While this decrease is visible among all degree levels, the higher the degree completed, the more likely the graduate has taken at least one quantitative course beyond Calculus II. The number of geoscience bachelor's graduates far exceed graduation rates for master's and doctoral degrees, so it is likely that the small number of bachelor's graduates that have taken the higher quantitative courses are the same students that move on to earn a graduate degree. Due to the complex nature of studying earth systems and geosciences, it is essential for recent graduates to have experience in coursework like Linear Algebra and Differential Equations to effectively understand areas such as fluid dynamics and systems modeling. The curriculum in geoscience departments may need to be reconsidered to include more emphasis on higher quantitative courses. When looking at the participation in quantitative courses by the type of institution, most of the participation in these higher quantitative courses occurred at doctorate granting institutions compared to liberal arts colleges, but some liberal arts institutions do offer higher level quantitative courses. If a large portion of graduate students with experience in courses, such as Differential Equations and Linear Algebra, gain this experience during their undergraduate degree, are liberal arts graduates starting at a disadvantage when applying for graduate school? Participation in Statistics courses has also raised some concern. Over the past three years, approximately 40 percent of geoscience doctoral graduates did not take a Statistics course during their postsecondary education. Participation in Statistics by doctoral graduates is expected to be much closer to 100 percent. It is essential for the geoscience academic workforce to have a solid background in Statistics in order to successfully interpret published research and add to the research within their fields.

Students were asked if they took an earth or environmental science course in high school and if they attended a two-year college for at least one semester before receiving their degree. From 2013-2016, approximately half the graduates took an earth or environmental science course in high school. In 2017, this trend continued, particularly among doctoral graduates with 69 percent of doctoral graduates taking an earth or environmental science course in high school. While these courses may or may not be the reason a student majors in the geosciences, high school exposure to earth science can create an interest in the subject area, as well as a comfort level in the subject entering into an introductory geoscience course in college.

Two-year college continues to be an important recruitment venue for the geosciences with nearly one third of bachelor's graduates and one quarter of master's graduates attending a two-year college for at least a semester. The percentage of doctoral graduates that attended a two-year college increased in 2017 compared to 2016 from 8 percent to 14 percent. Two-year colleges are becoming a viable and necessary option for many students to begin their postsecondary education.

Mary Lide Parker for AGI's 2017 Life as a Geoscientist contest

Elsemarie deVries, a Ph.D. candidate in the UNC Environmental Change Lab, obtains an aerial survey using a kite at Edisto State Park in South Carolina.

Quantitative skills and knowledge gained while working toward degree

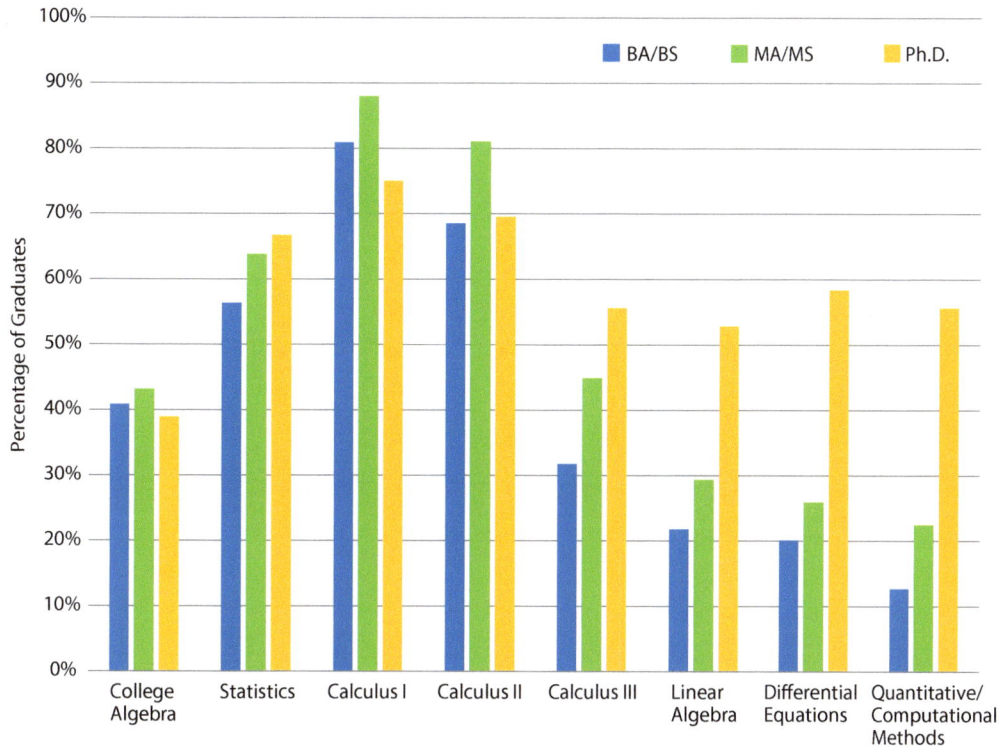

Quantitative skills and knowledge gained by graduates based on university classification**

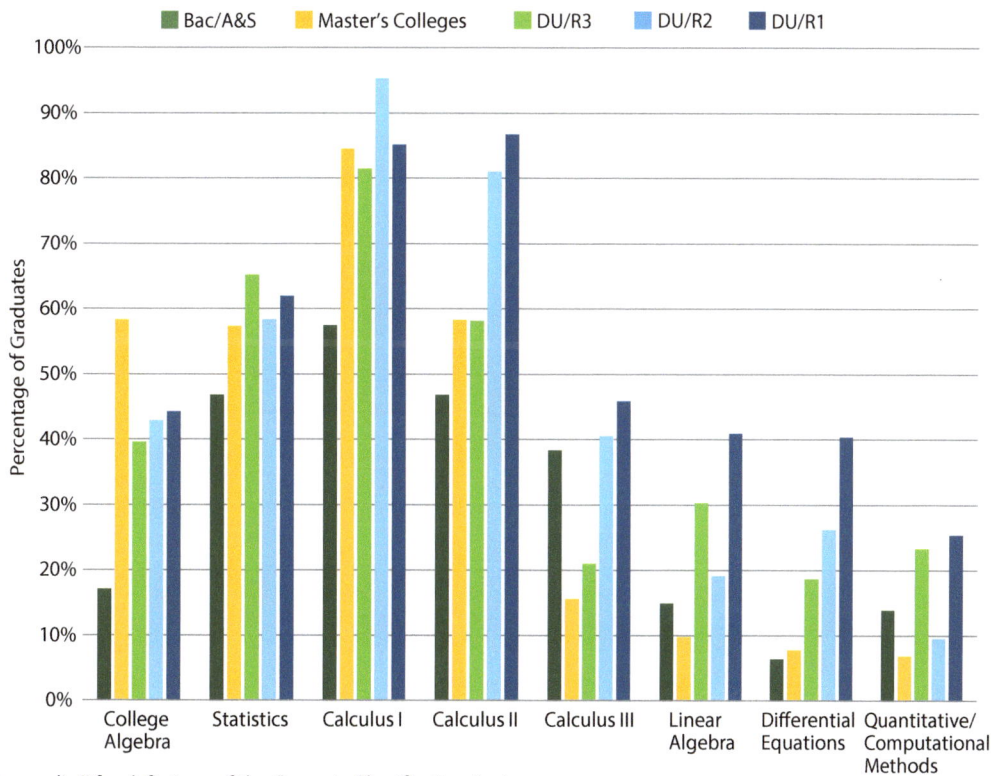

**See Appendix II for defintions of the Carnegie Classification System

Quantitative skills and knowledge gained while working toward degree by gender

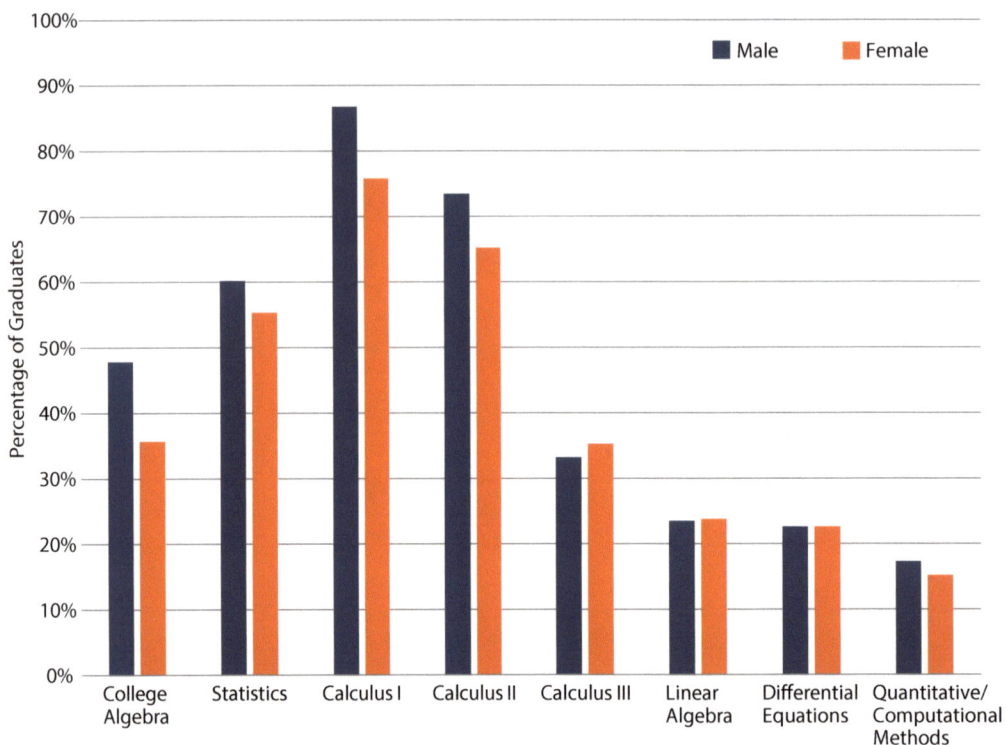

Percentage of graduates taking supplemental science courses

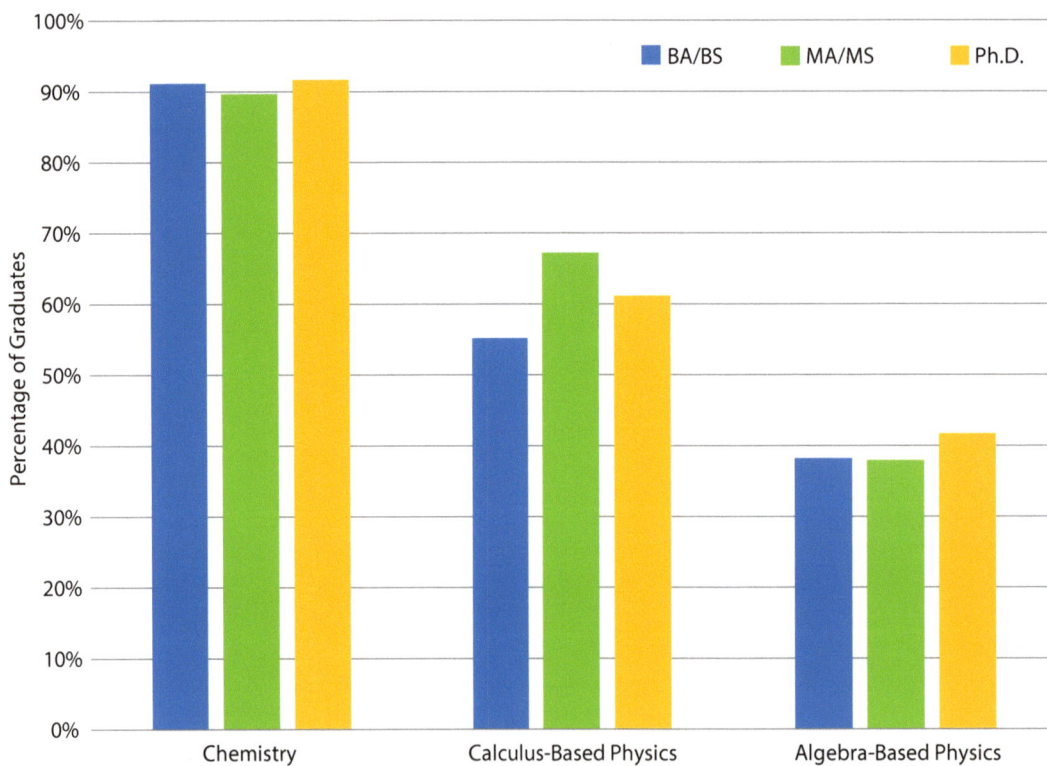

Graduates who took an earth science course in high school

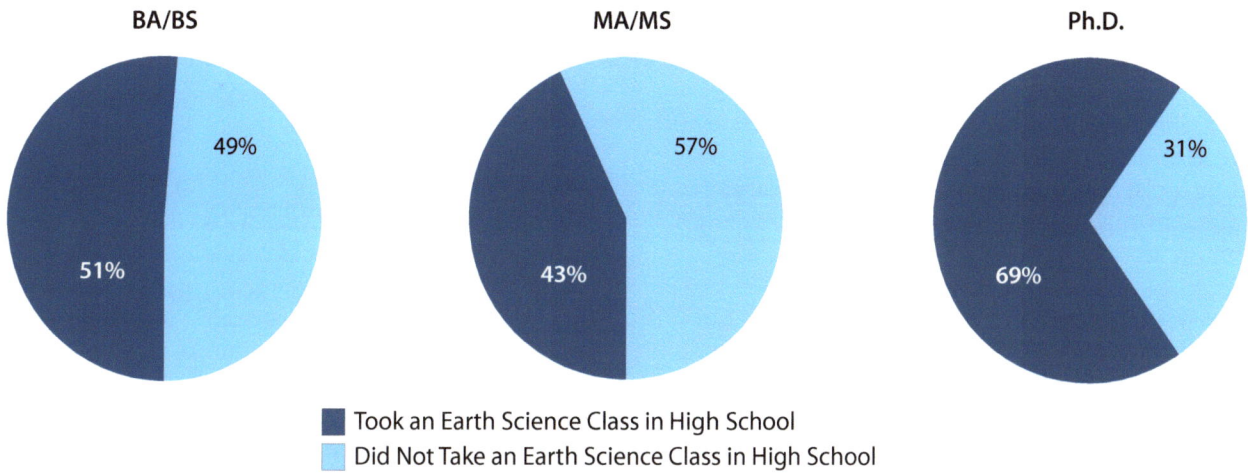

BA/BS	MA/MS	Ph.D.
49% / 51%	57% / 43%	31% / 69%

■ Took an Earth Science Class in High School
■ Did Not Take an Earth Science Class in High School

Graduates who attended a two-year college for at least 1 semester and took a geoscience course

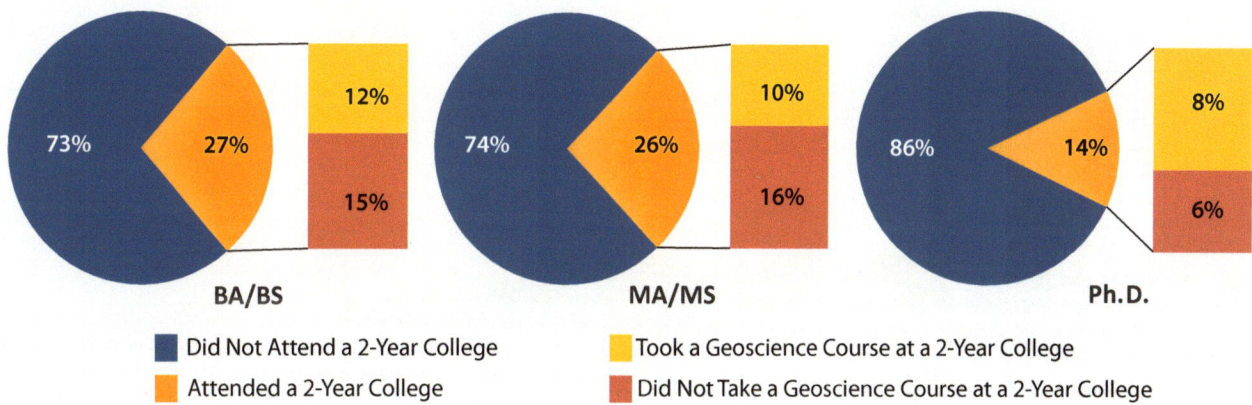

BA/BS	MA/MS	Ph.D.
73% / 27% — 12% / 15%	74% / 26% — 10% / 16%	86% / 14% — 8% / 6%

■ Did Not Attend a 2-Year College
■ Attended a 2-Year College
■ Took a Geoscience Course at a 2-Year College
■ Did Not Take a Geoscience Course at a 2-Year College

Choosing Geoscience as a Major

Graduates were asked which geoscience field they were pursuing, as well as the fields associated with any other postsecondary degrees. The chosen degree fields demonstrate the variety of disciplines related to the geosciences. Geology continues to be the most popular degree among undergraduates with students tending to specialize in different fields upon entering graduate school.

Consistently, the majority of graduates at the bachelor's and master's levels chose to major in the geoscience at some point during their undergraduate education. The timing of the choices of bachelor's and masters' students demonstrate the importance of the introductory geoscience courses for recruitment into the majors, but the doctoral graduates highlight the interdisciplinary nature of the geosciences with 11 percent of doctoral graduates changing their majors to the geoscience after receiving an undergraduate degree. Bachelor's or master's graduates in other sciences, particularly chemistry or physics, can easily transfer to the geosciences for future degrees with their strong physical science background. In 2017, the most doctoral graduates chose to major in the geosciences before beginning college, which was similar to the doctoral graduates in 2014 and 2015. This also goes along with the increase in doctoral graduates that took an earth or environmental science course in high school in 2017.

The graduates were asked to briefly explain their reasoning for majoring in the geosciences. As in previous years, the majority of graduates at all degree levels indicated the intellectual engagement of the geosciences as the reason for choosing their major. The comments included reasons related to the interdisciplinary nature of the geosciences, passion for certain degree fields, and the inherent interest of being outdoors and asking questions of the environment around them. In 2017, beyond reasons related to their interest in the subject area, bachelor's graduates also included reasons related to helping with societal issues related to earth science, career opportunities available to geoscientists, and the support provided by peers and faculty in their department. Master's graduates mentioned the career opportunities available to geoscientists. Doctoral graduates expressed their strong desire to continue research and field work in the geosciences and their ability to help with societal issues related to earth science.

Time when students decide to major in the geosciences

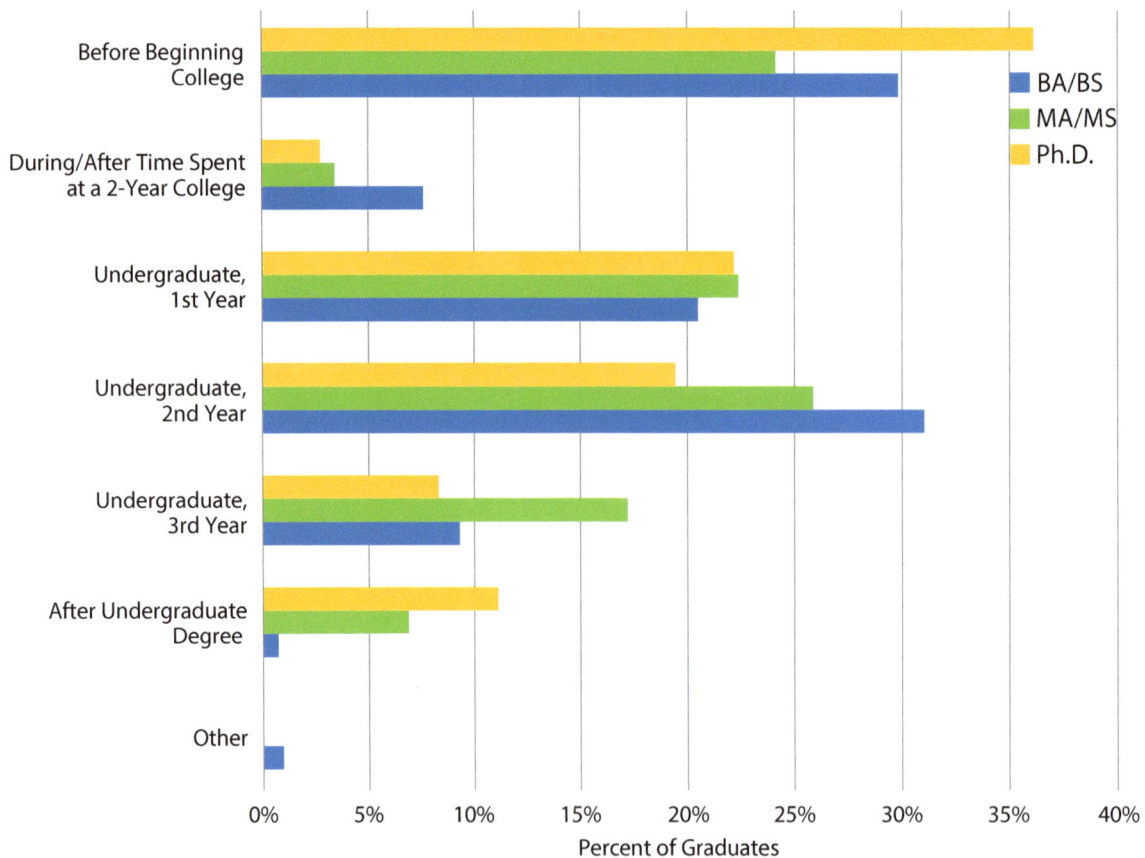

Chosen geoscience degree fields

Bachelor's Degree Graduates

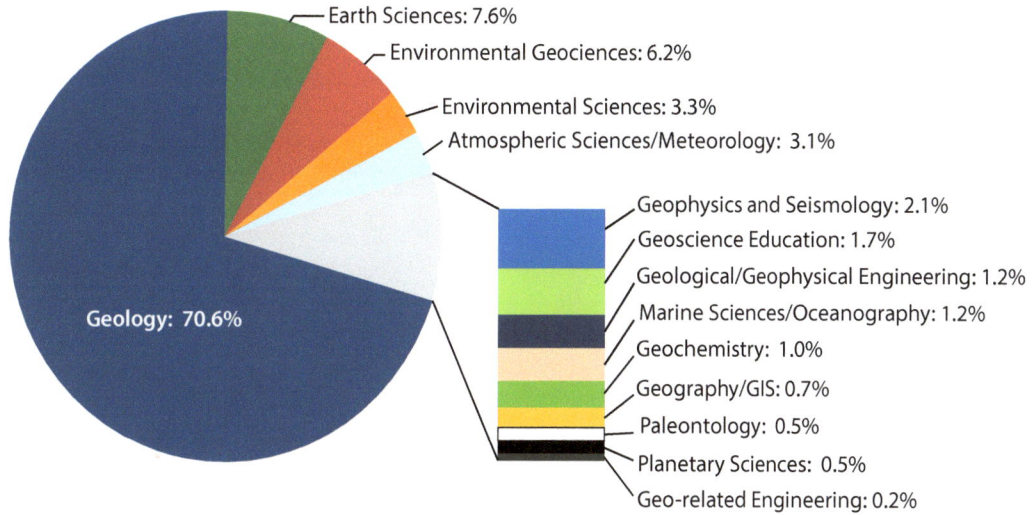

- Earth Sciences: 7.6%
- Environmental Geociences: 6.2%
- Environmental Sciences: 3.3%
- Atmospheric Sciences/Meteorology: 3.1%
- Geology: 70.6%
- Geophysics and Seismology: 2.1%
- Geoscience Education: 1.7%
- Geological/Geophysical Engineering: 1.2%
- Marine Sciences/Oceanography: 1.2%
- Geochemistry: 1.0%
- Geography/GIS: 0.7%
- Paleontology: 0.5%
- Planetary Sciences: 0.5%
- Geo-related Engineering: 0.2%

Master's Degree Graduates

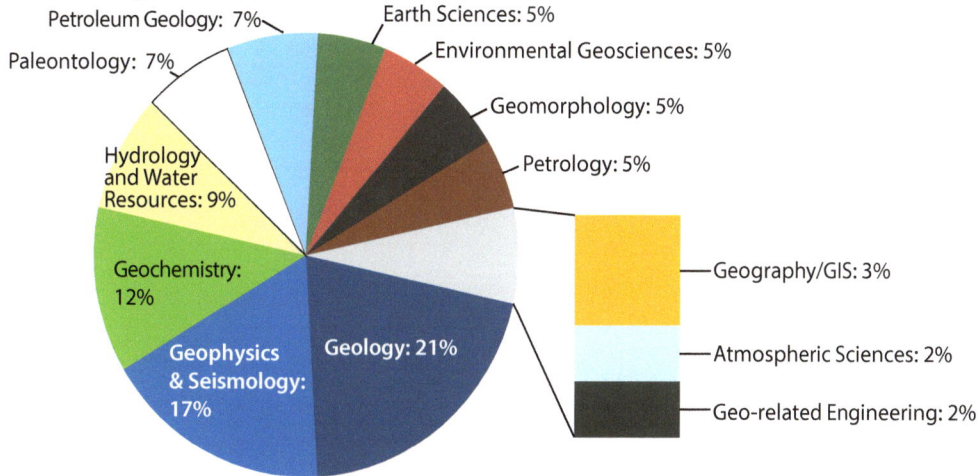

- Petroleum Geology: 7%
- Paleontology: 7%
- Hydrology and Water Resources: 9%
- Geochemistry: 12%
- Geophysics & Seismology: 17%
- Geology: 21%
- Earth Sciences: 5%
- Environmental Geociences: 5%
- Geomorphology: 5%
- Petrology: 5%
- Geography/GIS: 3%
- Atmospheric Sciences: 2%
- Geo-related Engineering: 2%

Doctoral Degree Graduates

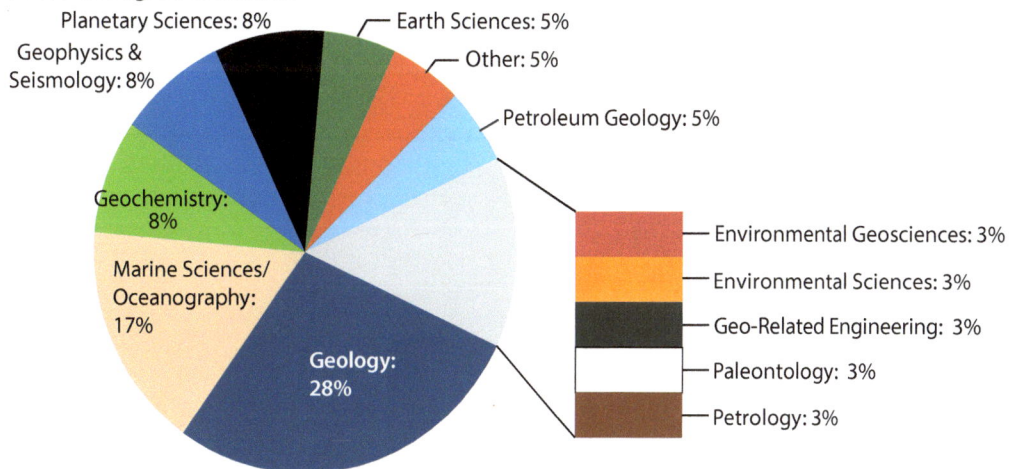

- Planetary Sciences: 8%
- Geophysics & Seismology: 8%
- Geochemistry: 8%
- Marine Sciences/Oceanography: 17%
- Geology: 28%
- Earth Sciences: 5%
- Other: 5%
- Petroleum Geology: 5%
- Environmental Geociences: 3%
- Environmental Sciences: 3%
- Geo-Related Engineering: 3%
- Paleontology: 3%
- Petrology: 3%

Ancillary Factors Supporting the Degree

Graduates were asked about their internship experiences while working towards their degree. Since the start of the Exit Survey project in 2013, there has been a consistent trend of low participation in internships, particularly among bachelor's and doctoral graduates with over half of them not participating in an internship. This trend did not change in 2017, but there was a small increase of 5 percent in the participation of bachelor's graduates in internship experiences. Master's graduates tend to have higher participation rates most likely due to their understanding of the importance of internship experiences to their professional career, but in 2015-2017, approximately 40 percent of master's graduates did not participate in an internship before graduation. To investigate these low internship participation rates further, the graduates are asked for the number of internship applications they submitted and the resources used to find internship announcements. In 2017, there was an increase in the percentages of graduates at all degree levels submitting at least one internship application while working on their degree compared to 2016. While more graduates did complete internship applications compared to last year, 37 percent of bachelor's graduates and 33 percent of doctoral graduates did not fill out an application.

The data suggest there might be two different issues related to the low participation rates in internships: the availability of internships to students and trouble finding such programs. Concern has been raised that there are not enough internship opportunities for geoscience students at all degree levels, and this is supported by the percentage of graduates that applied internships but were unable to secure one. Industry representatives have discussed the difficulties in providing these opportunities, such as the cost of an intern and the time spent training a temporary employee. The low participation rates may also be due to difficulty in finding appropriate opportunities than fit into an already packed degree program.

There is no centralized listing of internships for geoscience students, and departments may not be the best source for finding these opportunities. According to the graduates, most internship announcements were found through internet searches. It is essential that the geosciences community recognize the importance of internship activities for students' professional development. Consideration is needed for ways to promote and provide this type of professional development to current students at all degree levels.

Graduates were asked about their usage of financial aid while working on their degree. In 2017, 82 percent of bachelor's graduates, 81 percent of master's graduates, and 86 percent of doctoral graduates used at least one form of financial aid to complete their degree. Consistent with past years, in 2017, bachelor's graduates depended on student loans and federal grants to help pay for school, and master's and doctoral graduates tended to depend on research and teaching assistantships to help pay for school. There was an increase of approximately 20 percent of master's graduates that accepted a teaching assistantship among 2017 graduates compared to 2016 graduates. Master's graduates in 2017 also benefitted more from department scholarships and external scholarships compared to 2016. It is important to note that 38 percent of master's graduates and 14 percent of doctoral graduates used student loans to help pay for their degree. It is a bit of a misunderstanding that all graduate students have their degree paid for in full. Many schools may pay the tuition bill, but the attached fees are often up to the student to pay on their own. Also, due to increased pressure by universities to increase enrollments, some graduate programs can't afford to help financially support all of their students.

Graduates were also asked about their involvement with geoscience membership organizations. AGI is a federation of 52 geoscience societies, including the American Geophysical Union, the American Institute of Professional Geologists, the Geological Society of America, and the Society of Exploration Geophysicists. Professional societies can be useful tools for students and recent graduates to be successful as early-career geoscientists. While participation rates in geoscience organizations at all degree levels were relatively high, it appears that most students are not affiliated with one of the member societies in AGI's federation, which is surprising. However, participants in the survey can respond yes to their participation in geoscience organizations but choose not to list the organizations with which they are affiliated. Participation rates in AGI societies are likely higher than reported. Most students that do participate in a geoscience organization claim to be affiliated with a department level organization. This shows the importance of these department organizations to provide academic support and social connections for students in the program.

Number of internships held by graduating students

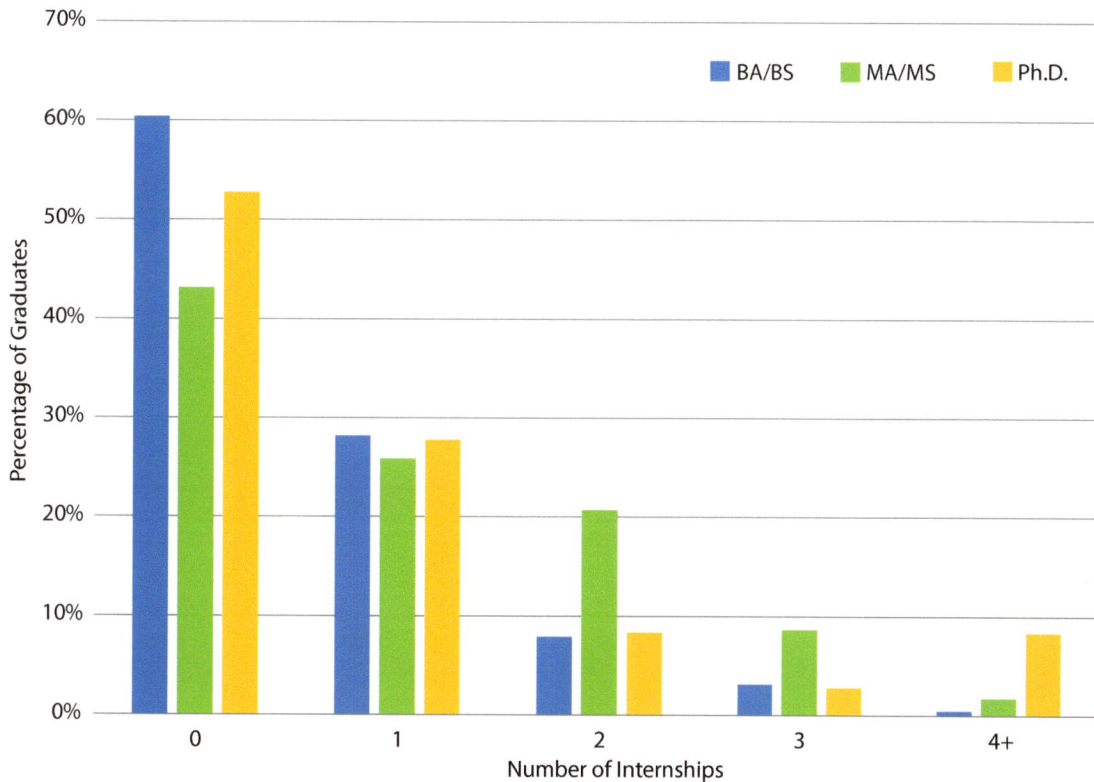

Internship applications completed by graduates

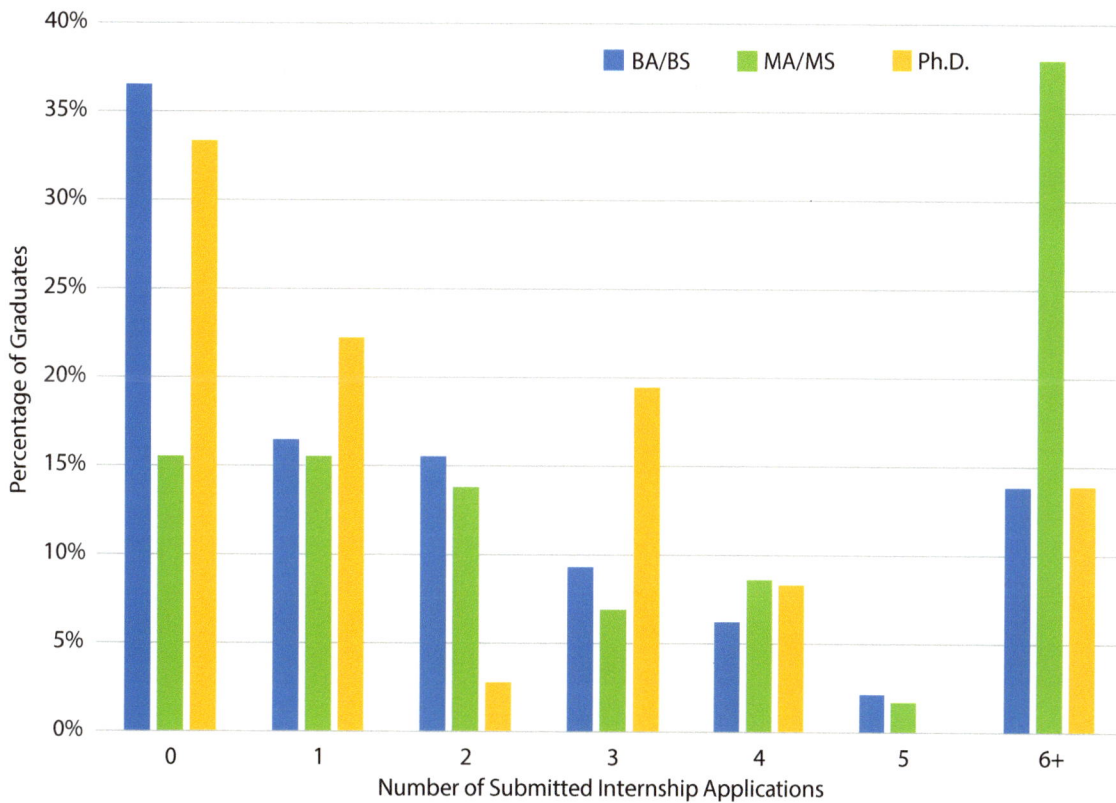

Resources used to find internship announcements

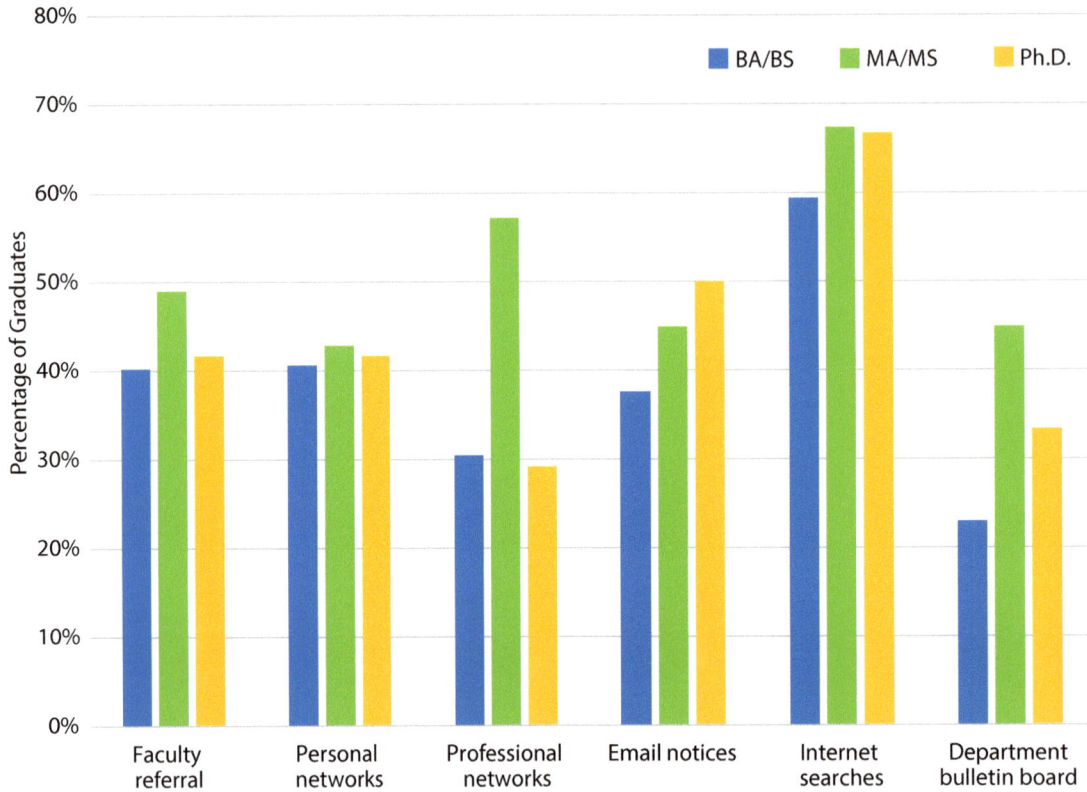

Types of financial aid used by graduating students while working towards a degree

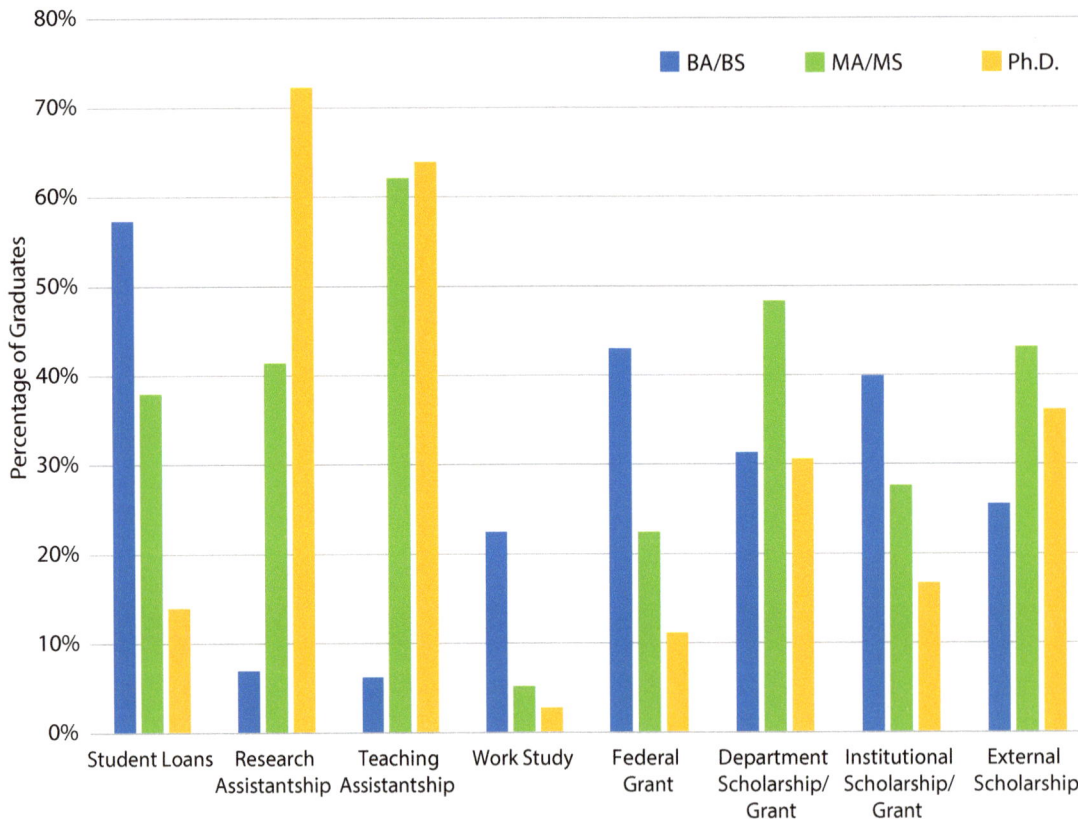

Participation in geoscience organizations

	BA/BS	MA/MS	Ph.D.
Associated with a geoscience-related club/organization	70%	74%	64%
Participated in department-level geoscience club	46%	53%	39%
Member of an AGI Member Society	25%	52%	31%
Member of an Honor Society	11%	7%	6%

Average GPA

	BA/BS	MA/MS	Ph.D.
Average years to degree completion	3.17	3.52	5.03
Average overall GPA	3.32	3.79	3.84
Average geoscience GPA	3.45	3.83	3.84

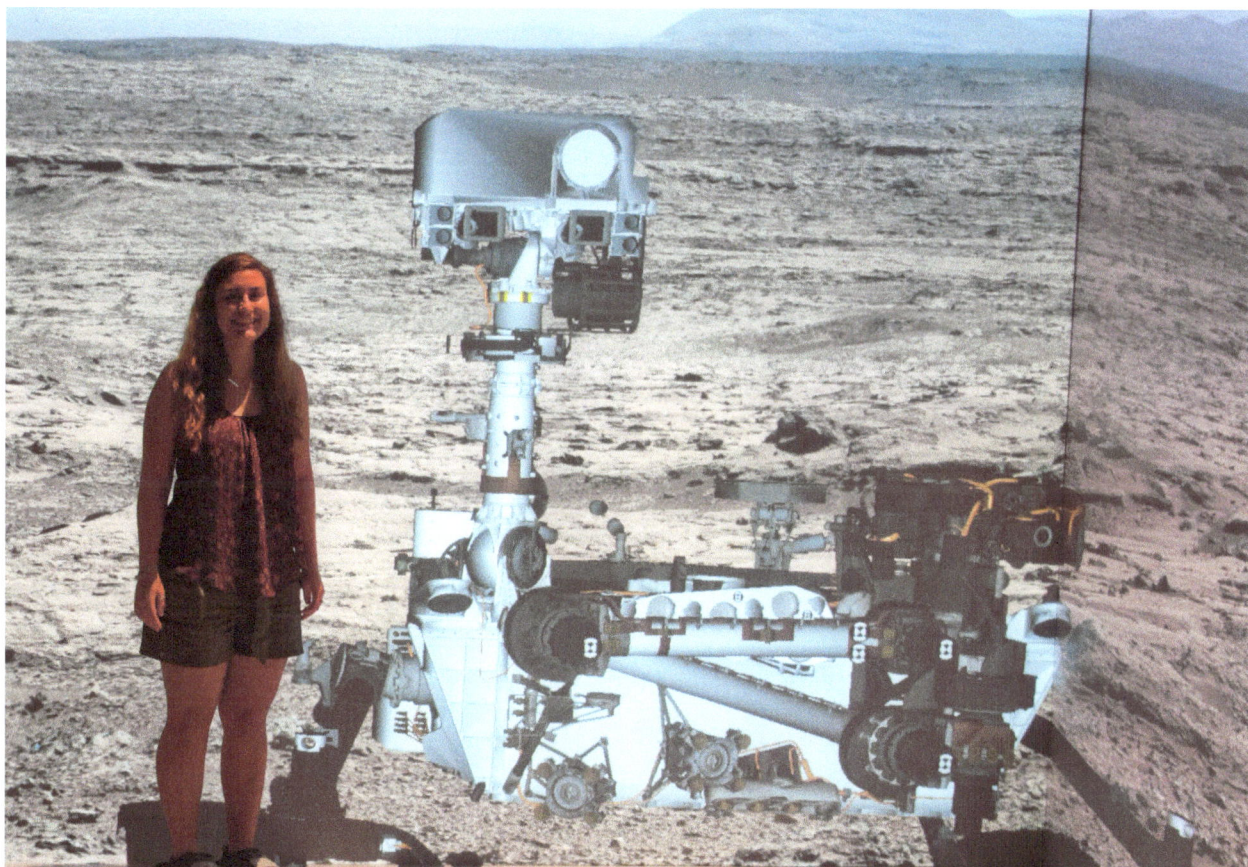

Schuyler Borges for AGI's 2017 Life as a Geoscientist contest

Photo shoot with the Curiosity rover.

Field Experiences

Clear definitions were set to distinguish between field camp, field courses, and field experiences. A field camp was defined as an academic program lasting four or more weeks that is primarily focused on field tools and methods. Because field camp is typically an experience only taken once, this question covers the graduates' entire postsecondary education. A field course was defined as a course with a field component primarily covering field methods and experimentation that utilized at least half of the total class time. A field experience was defined as any course that contained a field component, such as a field trip, field work, or other time in the field, that is not included in the definitions for field camp or field courses.

In 2017, approximately 1 percent of geoscience graduates did not participate in any field experiences while working on their degree. Since 2014, there has been a 13 percent drop in the participation rate of doctoral graduates participating in a field camp experience with 47 percent participation in 2017. While field camp participation rates among bachelor's and master's graduates have stayed relatively the same from 2014-2016, in 2017, there was an increase in participation among graduates at both degree levels. Participation in field camp among bachelor's graduates increased by 7 percent in 2017 compared to 2016 and participation among master's graduates increased by 10 percent in 2017 compared to 2016. Participation in field courses and field experiences increased among graduates at all degree levels. Field experiences have become a necessary part of the curriculum, and most employers look for these experiences among job applicants. Participation in field experiences was high regardless of the type of institution. However, availability of field camp opportunities continue to be low among liberal arts colleges.

Field experiences, particularly those that teach effective field skills, for geoscience students are essential for their education and training for the workforce. Since 1996, participation rates in field camp have soared to a point where many field camps are at capacity each year. Because of this overall increased participation in field camp across the United States, many employers expect recent graduates to have the necessary field skills developed through a field camp course. For those recent graduates that do not have access to a field camp, many of those necessary skills can be gained through field courses, as long as participation in these courses is encouraged by the geoscience programs.

Student participation in field experiences based on university classification**

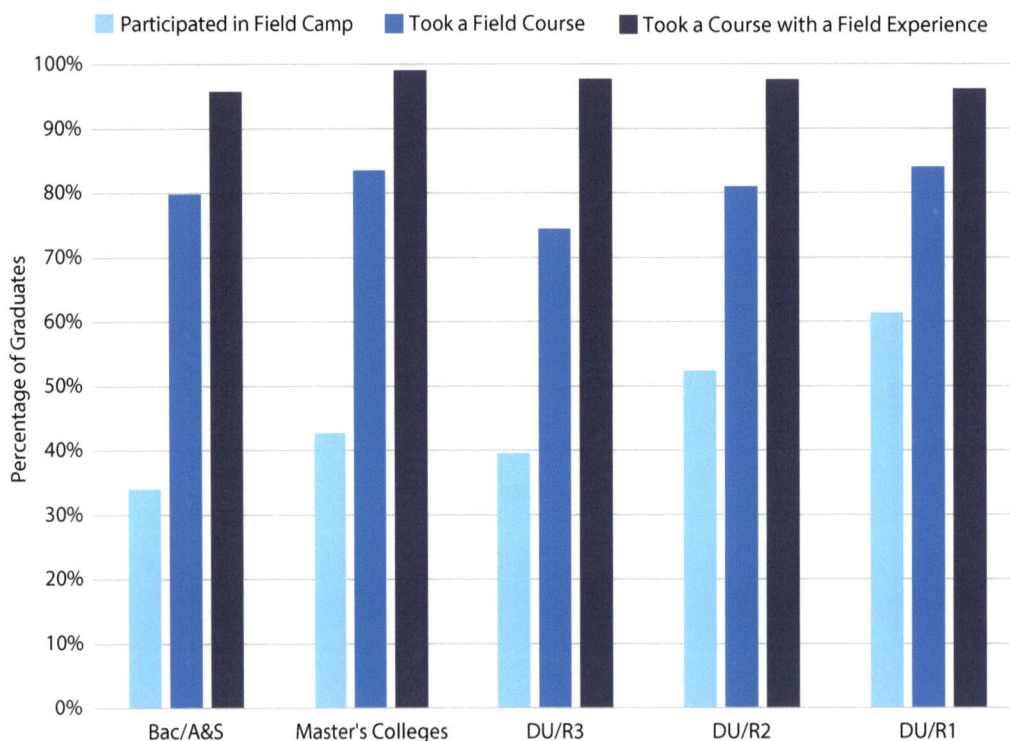

**See Appendix II for definitions of the Carnegie University Classification System

Graduating students who have participated in field camp

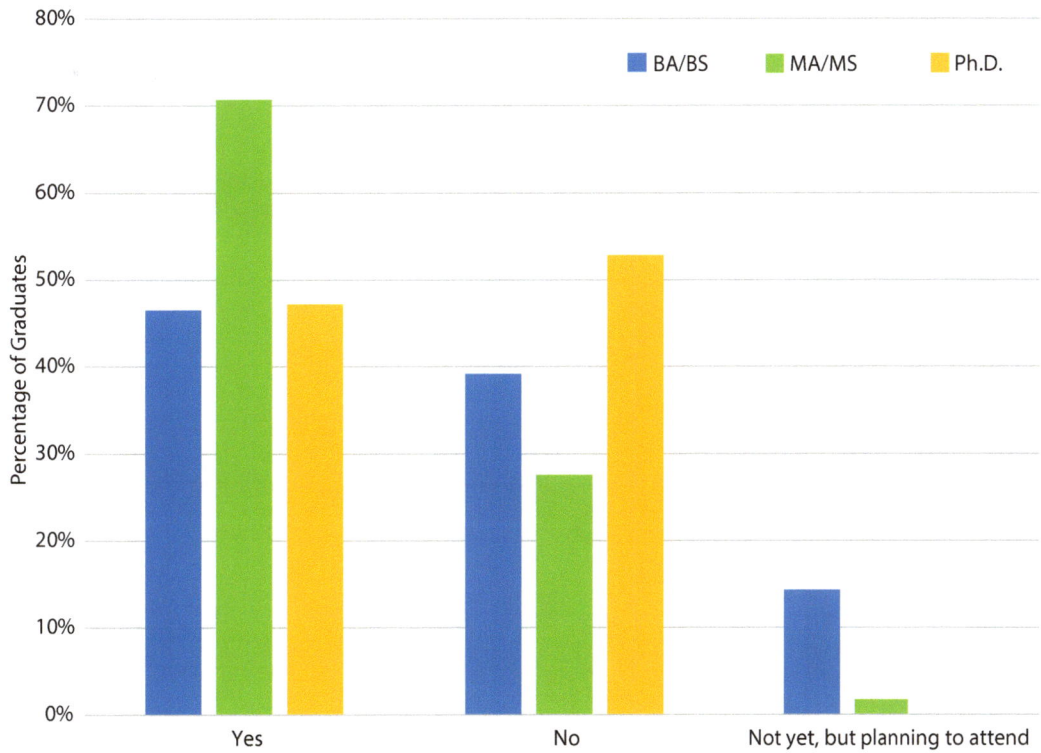

Graduating students who have participated in field camp by gender

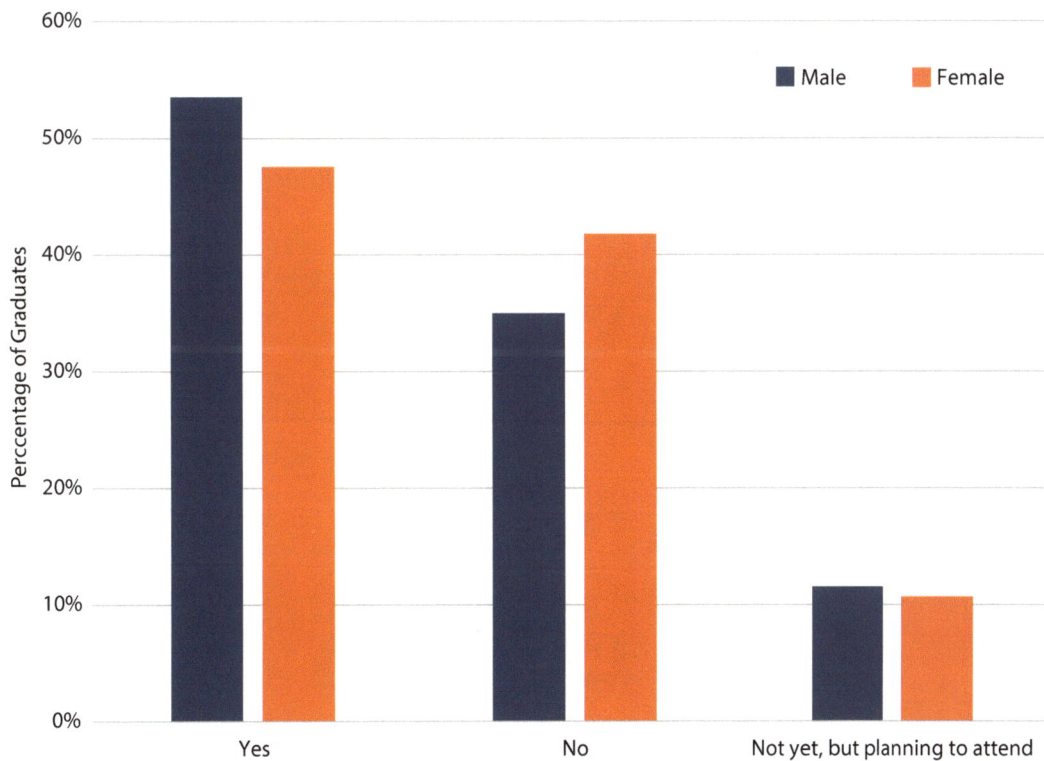

21

Graduates who have taken one or more field courses

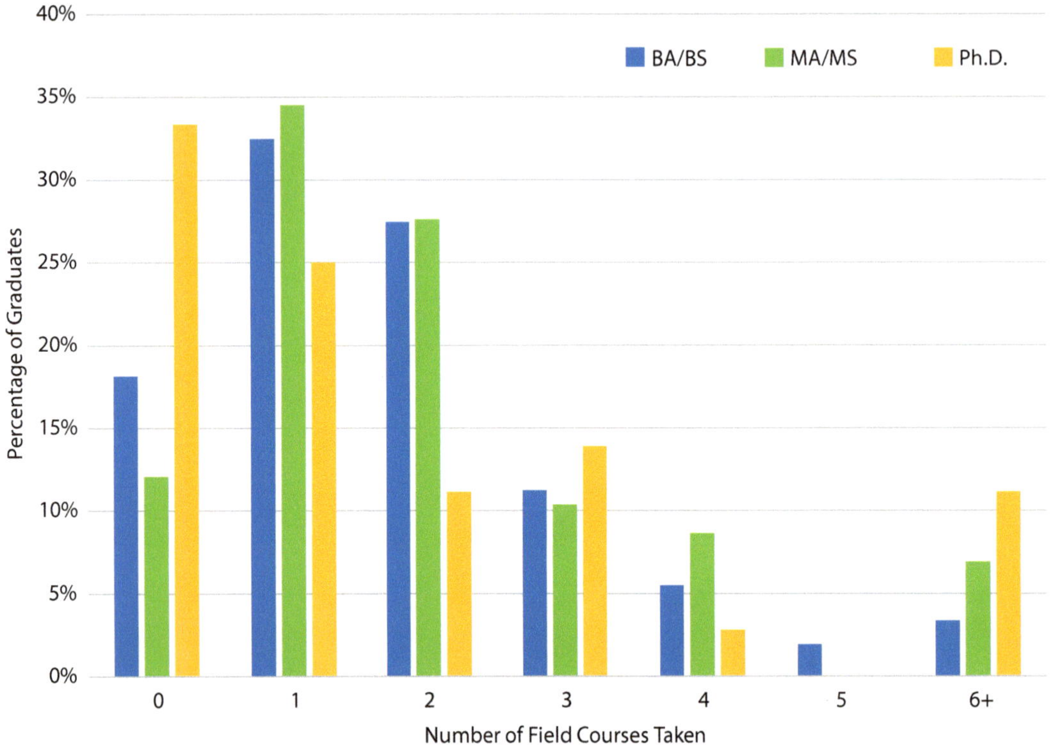
40%
35%
30%
25%
20%
15%
10%
5%
0%
Percentage of Graduates
BA/BS
MA/MS
Ph.D.
0
1
2
3
4
5
6+
Number of Field Courses Taken

Graduates who have taken one or more courses with a field experience

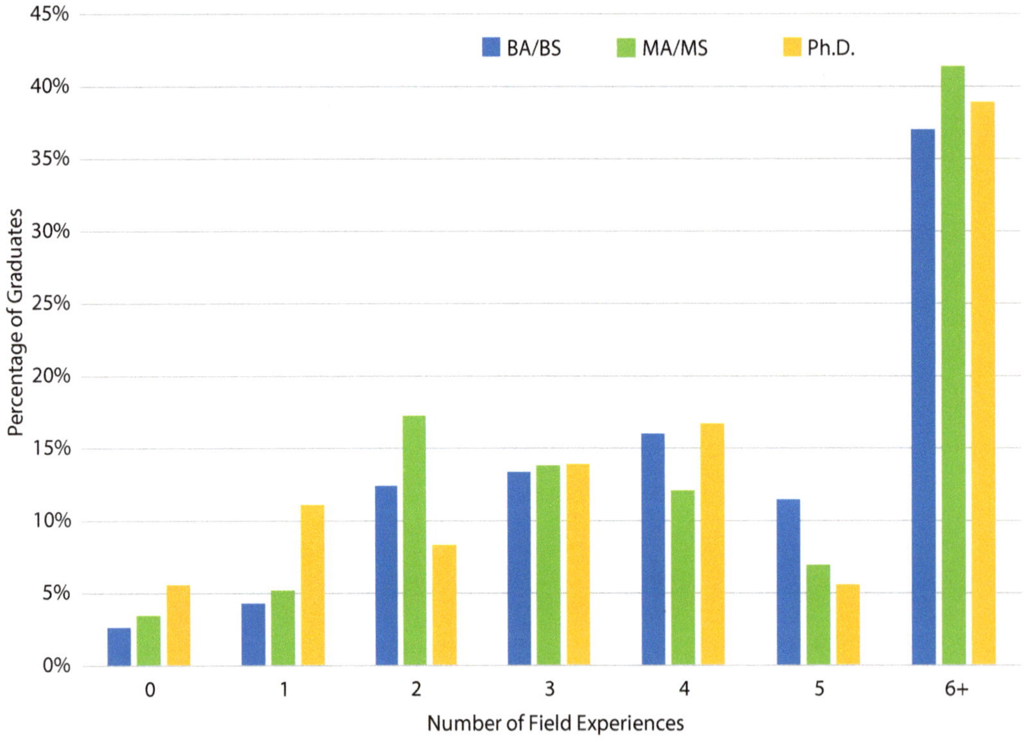
45%
40%
35%
30%
25%
20%
15%
10%
5%
0%
Percentage of Graduates
BA/BS
MA/MS
Ph.D.
0
1
2
3
4
5
6+
Number of Field Experiences

Lawrence Malinconico for AGI's 2017 Life as a Geoscientist contest

Lafayette College geology majors Emilie Henry and Charles Chrin mapping in the Bighorn Basin, Wyoming. A mix of old-school and modern techniques: traditional Brunton compass and digital recording and mapping on an iPad.

Research Experiences

The graduates were asked about their research experiences while working towards their degrees. If they indicated participation in at least one research experience, the graduates were then asked about their participation in faculty-directed research and self-directed research. If they indicated participation in self-directed research, they were asked to identify the basic research methodology used to conduct their research.

In 2017, 20 percent of bachelor's graduates and 5 percent of master's graduates did not participate in a research experience before completing their degree. That is a 5 percent increase in master's graduates participating in at least one research experience compared to 2016.

In 2017, there was a decrease in undergraduate students using literature-based methods by 12 percent, field-based methods by 5 percent, and computer-based methods by 5 percent for conducting their self-directed research compared to 2016. Among graduate students, there was a 7 percent decrease in the use of field-based methods, a 10 percent increase in the use of lab-based methods, and a 15 percent decrease in the use of computer-based

methods for their self-directed research. The decrease among graduate students using computer-based methods is surprising because in previous years that percentage had been increasing, and computer modeling has become a major tool for geoscience research in recent years. However, there were fewer graduate students that participated in the survey this year working in the field of atmospheric sciences and meteorology, which could explain some of that decrease. Interestingly, more female graduate students used computer-based methods and literature-based methods for their self-directed research than males.

When asked about the importance of research experiences to the graduates' academic and professional development, 81 percent of bachelor's graduates, 85 percent of master's graduates and 97 percent of doctoral graduates rated these experiences as "very important". Outside of the classroom, these research experiences provide one of the best opportunities for students to utilize their critical thinking skills and learn how to work with uncertainty and imperfect data sets. It is encouraging to see high participation rates since the start of AGI's Exit Survey in research experiences at all degree levels.

Research methods utilized by graduates in their self-directed research

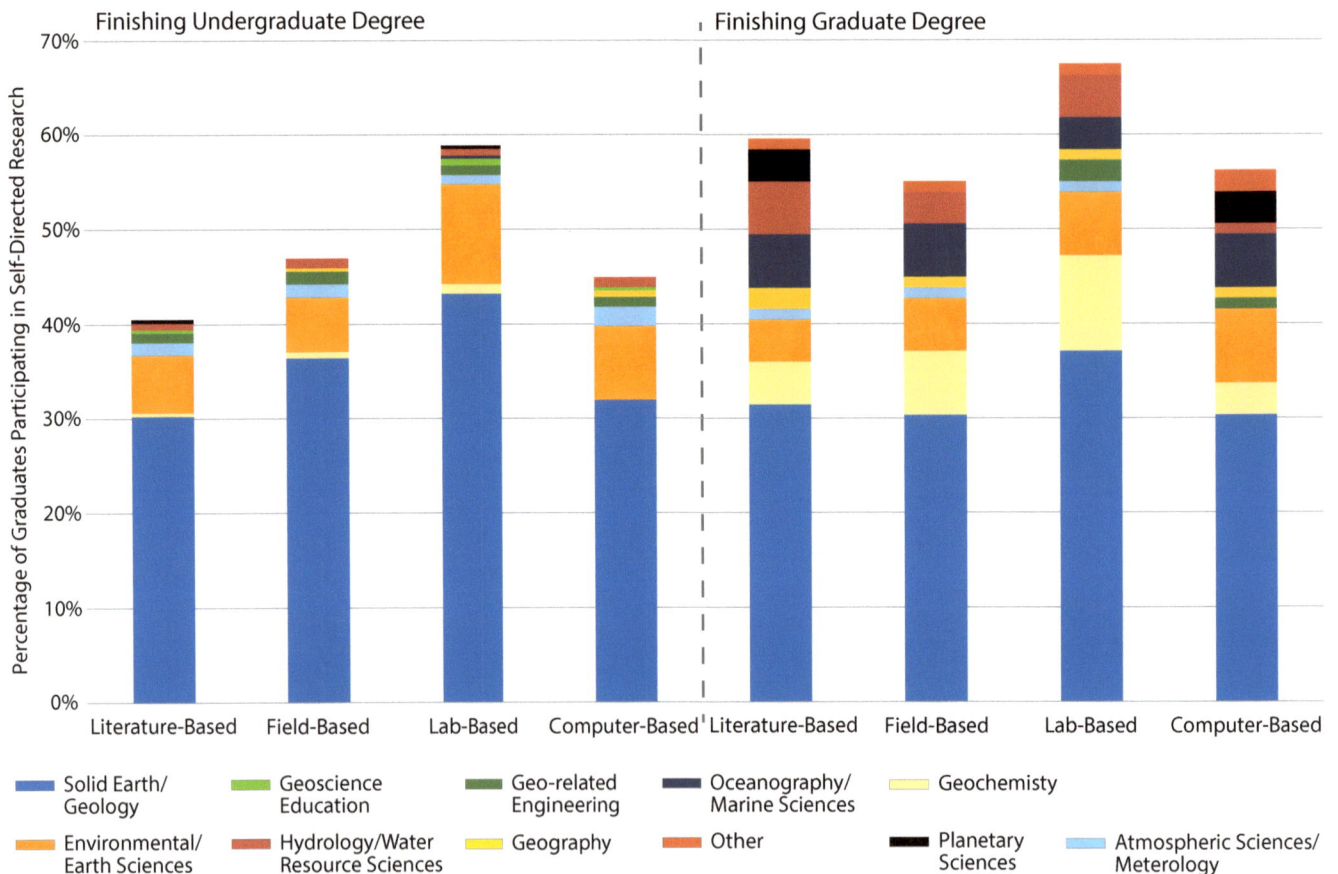

Participation rate of graduates in research experiences

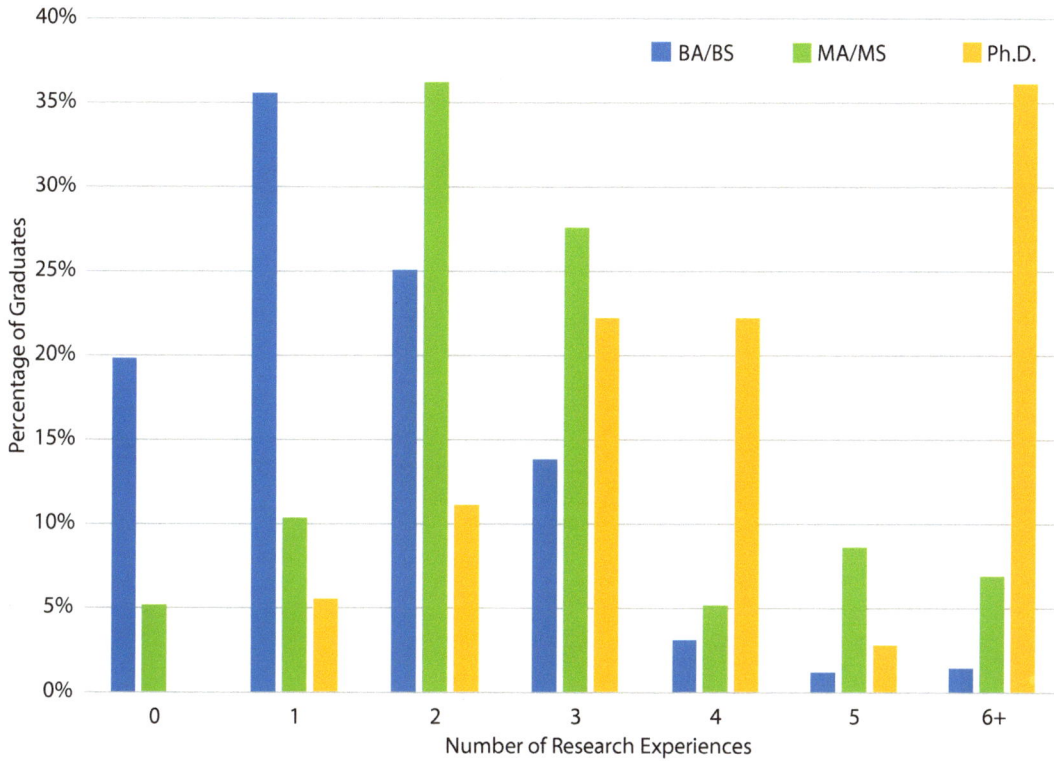

Student participation in faculty-directed and self-directed research

	BA/BS	MA/MS	Ph.D.
Faculty-Directed Research	58%	78%	81%
Self-Directed Research	70%	91%	100%

Research methods utilized by graduates in their self-directed research by gender

Finishing Undergraduate Degree | Finishing Graduate Degree

■ Male   ■ Female

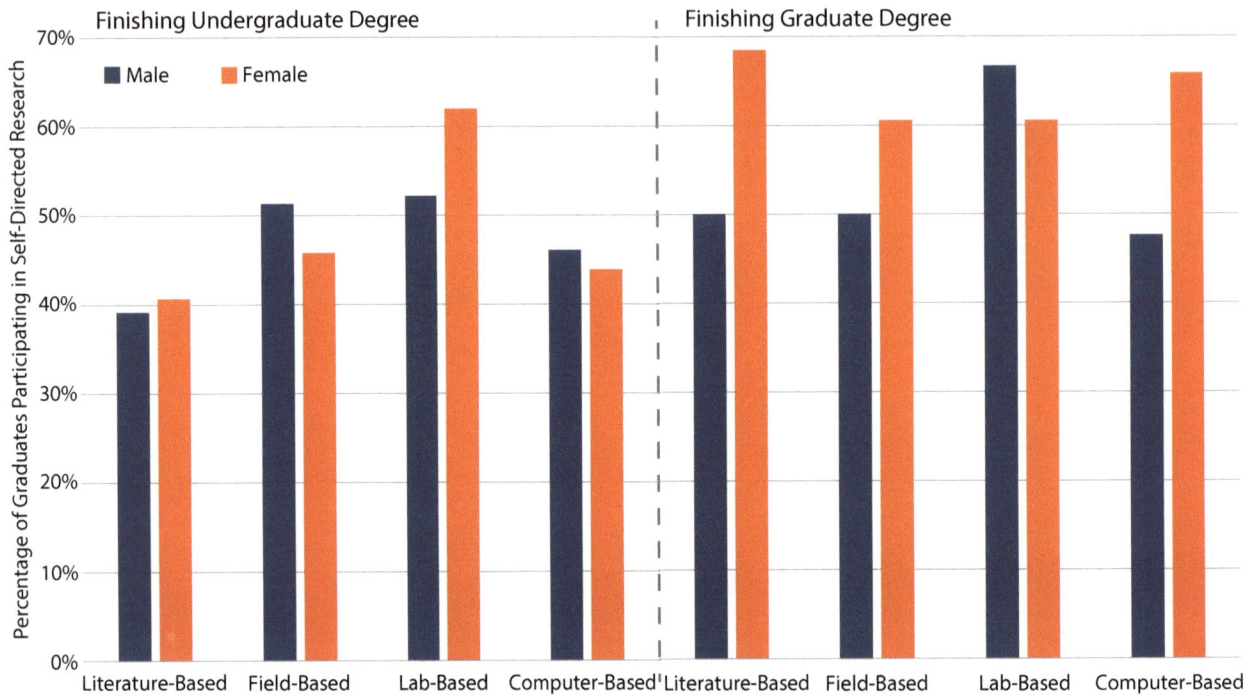

Y-axis: Percentage of Graduates Participating in Self-Directed Research (0% – 70%)

X-axis categories: Literature-Based, Field-Based, Lab-Based, Computer-Based | Literature-Based, Field-Based, Lab-Based, Computer-Based

Student participation in research based on university classification**

■ No Research Experience   ■ Collaborated with Faculty   ■ Conducted Individual Research

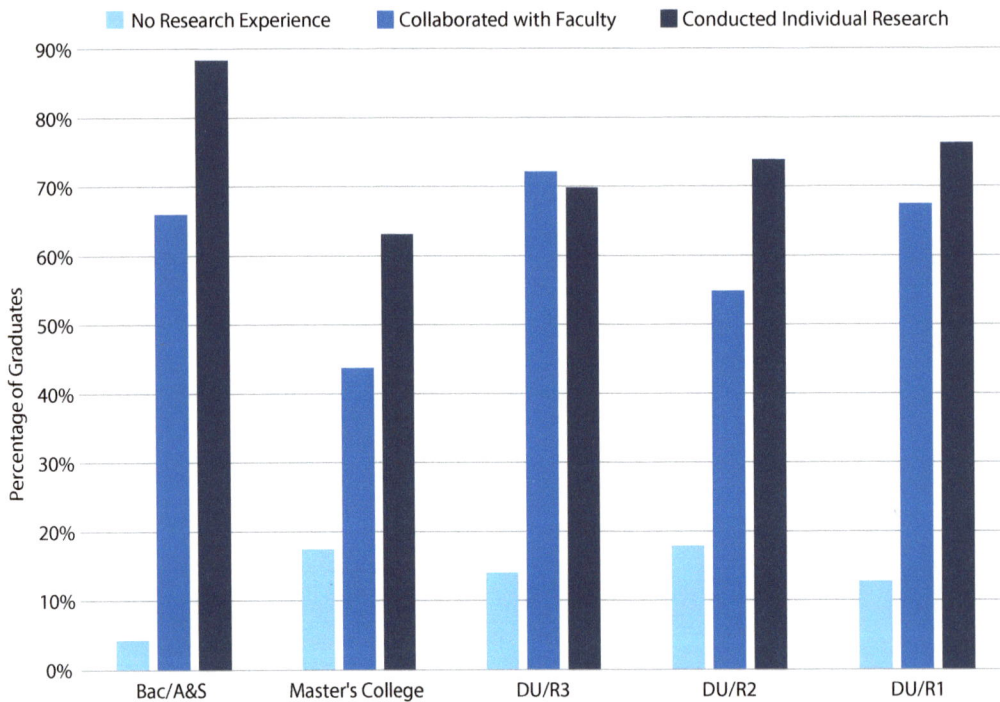

Y-axis: Percentage of Graduates (0% – 90%)

X-axis categories: Bac/A&S, Master's College, DU/R3, DU/R2, DU/R1

**See Appendix II for definitions of the Carnegie University Classification System

Rob Thomas for AGI's 2017 Life as a Geoscientist contest

A Sherpa mountain guide for the Khumbu Climbing Center in Phortse, Nepal, taking bearings off of peaks around Phortse in order to located himself on a map. These skills, along with knowledge of how to recognize and avoid natural hazards, weather issues and the impacts of climate change, especially on the glaciers, is helping to improve the safety of mountain workers in Nepal.

Future Plans: Working Toward a Graduate Degree

The graduates were asked if they have immediate plans to continue their education. Those indicating plans for a graduate degree after graduation were then asked to share the degree they would pursue and the field of interest for the degree.

From 2014-2017, the percentage of bachelor's graduates immediately planning to attend graduate school has ranged from 42 percent to 35 percent, and the percentage of master's graduates immediately planning to work towards another graduate degree has ranged from 20 percent to 31 percent. Interestingly, it was the bachelor's graduates in 2017 that reached the lowest point of 35 percent interested in graduate school, and the master's graduates in 2017 reached the highest point of 31 percent interested in graduate school. This may be indicative of other trends seen in the geosciences. AGI's annual enrollment data updates have seen relatively stagnant growth in the enrollments in geoscience graduate programs, and conversations with departments have indicated that many of these graduate programs are at capacity, which has created competition for the open slots in these programs. Bachelor's graduates recognize the need to start earning money to pay back student loans. However, master's graduates have spent the past two to three years in the graduate academia network, which may encourage these graduates to attempt another graduate degree and gain more skills while waiting for hiring in particular industries to pick up

While the majority of bachelor's and master's graduates planning to attend graduate school were interested in a wide array of geoscience degree fields, there were a few recent graduates planning to obtain a graduate degree in a non-geoscience field, such as physics, biological sciences, law, medical sciences, and business. It's not clear if these recent graduates plan to use their geoscience knowledge in conjunction with these other degree fields or if there students are planning to move away from the geoscience workforce.

students planning to attend graduate school after graduation

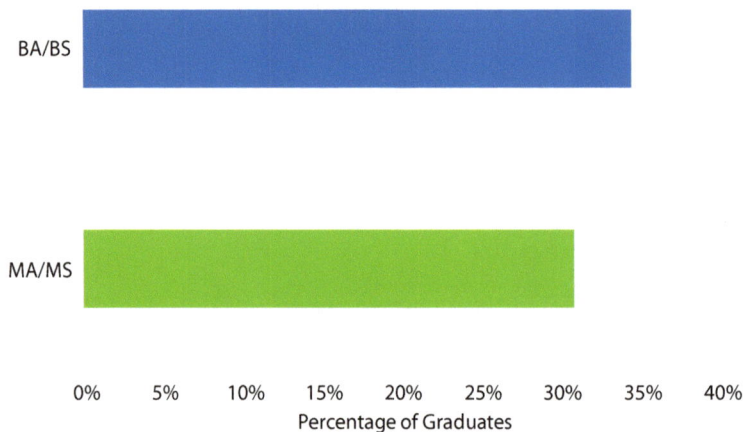

Students planning to attend graduate school after graduation by gender

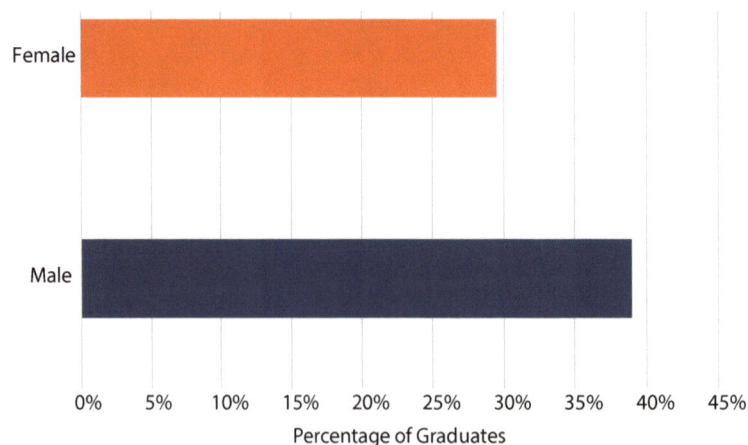

Students graduating with an undergraduate degree

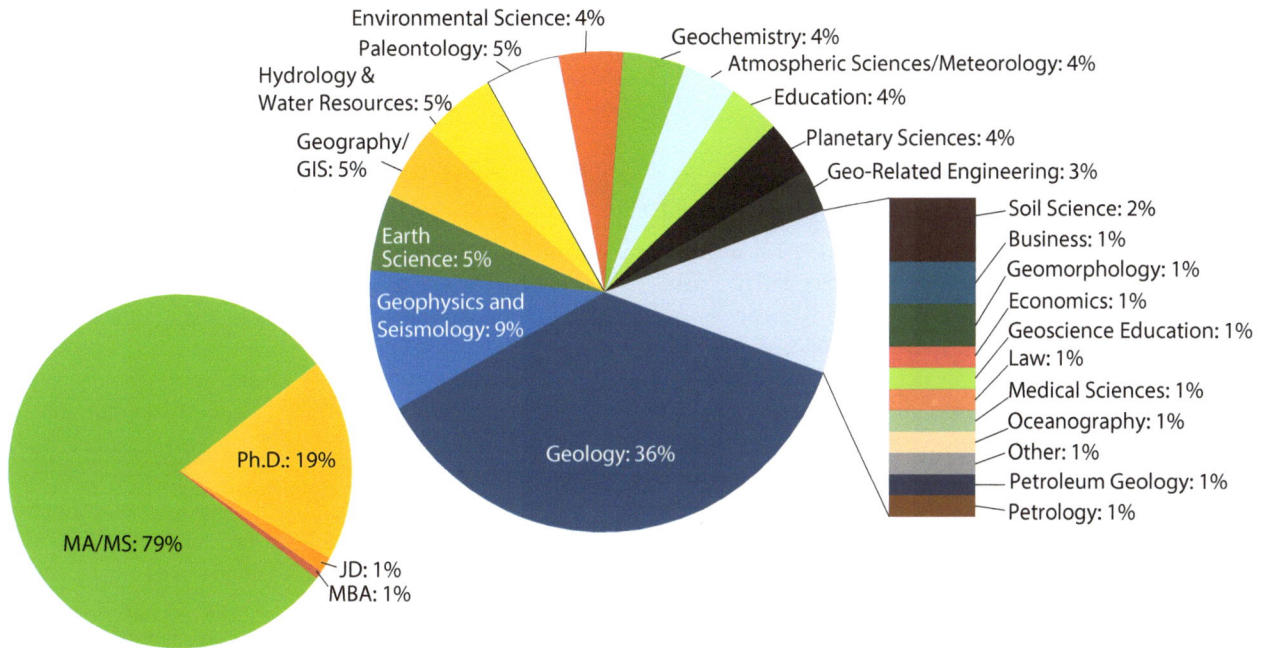

Possible Future Graduate Degree

MA/MS: 79%
Ph.D.: 19%
JD: 1%
MBA: 1%

Possible Future Field of Study

Environmental Science: 4%
Paleontology: 5%
Hydrology & Water Resources: 5%
Geography/GIS: 5%
Earth Science: 5%
Geophysics and Seismology: 9%
Geochemistry: 4%
Atmospheric Sciences/Meteorology: 4%
Education: 4%
Planetary Sciences: 4%
Geo-Related Engineering: 3%
Geology: 36%

Soil Science: 2%
Business: 1%
Geomorphology: 1%
Economics: 1%
Geoscience Education: 1%
Law: 1%
Medical Sciences: 1%
Oceanography: 1%
Other: 1%
Petroleum Geology: 1%
Petrology: 1%

Students graduating with a graduate degree

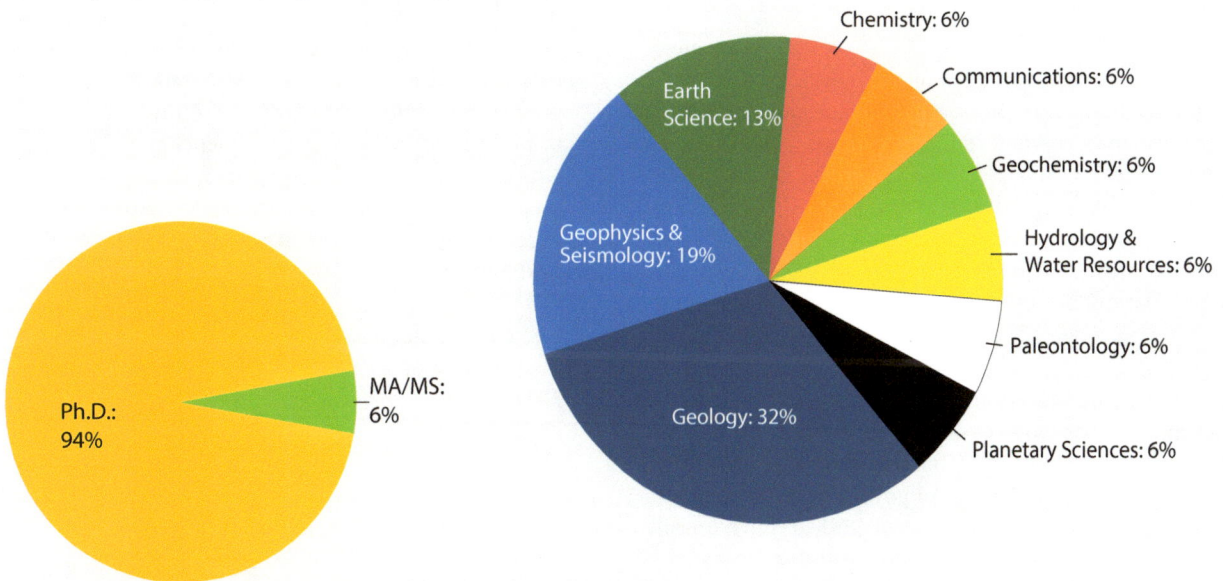

Possible Future Graduate Degree

Ph.D.: 94%
MA/MS: 6%

Possible Future Field of Study

Chemistry: 6%
Communications: 6%
Geochemistry: 6%
Hydrology & Water Resources: 6%
Paleontology: 6%
Planetary Sciences: 6%
Geology: 32%
Geophysics & Seismology: 19%
Earth Science: 13%

Future Plans: Working in the Geosciences

The graduates were asked if they had accepted or were seeking a job position within the geoscience workforce. If they had accepted a job, they were asked questions about these accepted job positions. Because the graduates take the survey right around graduation, it is not surprising that there are still relatively high percentages of graduates at all degree levels still seeking employment. However, this year, hiring of graduates at all degree levels was at its lowest in five years, particularly among doctoral graduates. This downward trend has been seen nearly every year since starting this survey. In 2017, 10 percent of bachelor's graduates, 31 percent of master's graduates, and 36 percent of doctoral graduates had secured a job in the geosciences at the time of graduation. Since 2014, doctoral graduates that found a geoscience job have dropped from 70 percent to 36 percent, and the percentage of master's graduates that found a geoscience job was at its lowest in four years in 2017. Substantial structural changes are underway in the geoscience job market, with machine learning processes replacing many basic geoscience jobs, coupled with the slow recovery in energy and mining. Additionally, with the rollback of many environmental regulations, hiring in the environmental and engineering sector appears to be cautious at the moment.

Among those graduates that were able to find a geoscience job, the bachelor's graduates tend to find positions in more industries than master's or doctoral graduates, and the jobs found by the master's and doctoral graduates tend to be in more traditional geoscience industries. The environmental services industry continues to be viable for bachelor's graduates with 30 percent of bachelor's graduates that found a job entering into that industry. Over the past four years, the oil and gas industry tended to dominate the hiring of master's graduates at the point of graduation, but in 2017, there was a significant drop in the percentage of master's graduates hired by the oil and gas industry from 60 percent in 2016 to 28 percent in 2017. This year also saw in increase in the master's graduates hired by the federal government from 12 percent in 2016 to 28 percent in 2017, but no doctoral graduates in 2017 were hired by the federal government at the point of graduation. Among the doctoral graduates that found a job, 69 percent are working as a postdoc, and these postdocs are all at 4-year universities or research institutes.

As in previous years, the annual salaries for the geoscience jobs secured by the 2016 graduates show ranges by degree level, but these ranges are not as defined as in past years. In previous years, master's graduates tended to fall within two different ranges depending on the industry that hired them, with the higher salaries awarded by the oil and gas industry. However, with the lower hiring of master's graduates in 2017 in the oil and gas industry, the starting salary range for master's graduates does not reach above $100,000, as in past years, and covers the range of bachelors' graduate starting salaries and most of the doctoral graduates' starting salaries.

Graduates that found geoscience employment were asked to identify the resources they used to find their job. Since 2014, graduates from all degree levels have noted the use of personal contacts as a major resource for finding their job. Faculty referrals tend to be a useful resource for graduates at all degree levels, particularly among bachelor's graduates, which supports the importance of the faculty in helping their students move forward after graduation. Beyond personal contacts, master's graduates in past year have consistently relied on the campus recruitment events for finding a job. However, in 2017, master's graduates depended much more on internet searches than any other resource. With the decrease in hiring in the oil and gas industry among master's graduates, these campus recruitment events must be dominated by energy companies. It also raises the question if the organizers of these events need to diversify the industry representation at the campus recruitment events. This year, conference networking became a very useful resource for masters' graduates and doctoral graduates to find jobs compared to past years. Many of the geoscience member societies with student membership have placed a stronger emphasis on preparing students for entering the workforce, so it is encouraging to see these efforts may be helping students find jobs.

The circular figure displays the connection between the degree fields of recent geoscience graduates from 2013-2017 (in color) to the industries where geoscientists found their first job after graduation (in gray). The size of the bars along the outer edge of the circle represents the number of recent graduates that pursued a particular degree field and entered a particular industry. Each colored, inner ribbon connects a particular degree field with a job in a particular industry. The visualization shows the variety of industries available to graduates with a geoscience degree, as well as the complexity of the workforce and knowledge needed in the distinct industries.

Industries where graduating students have accepted a job within the geosciences

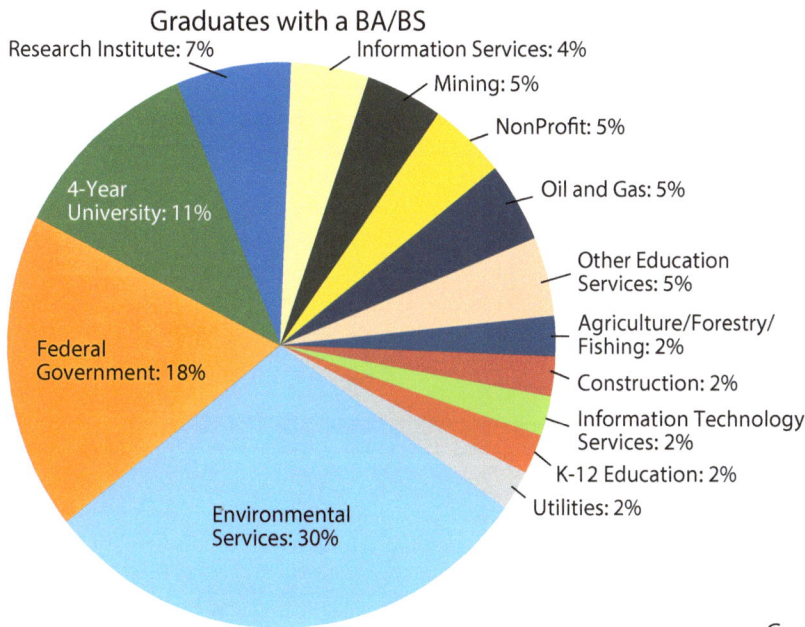

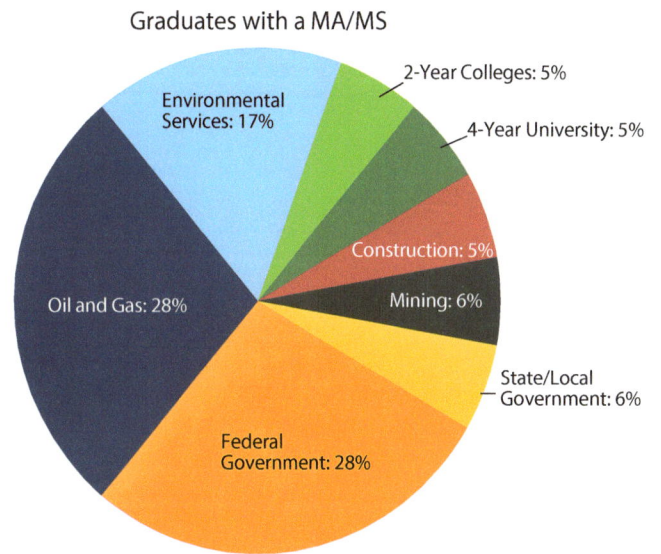

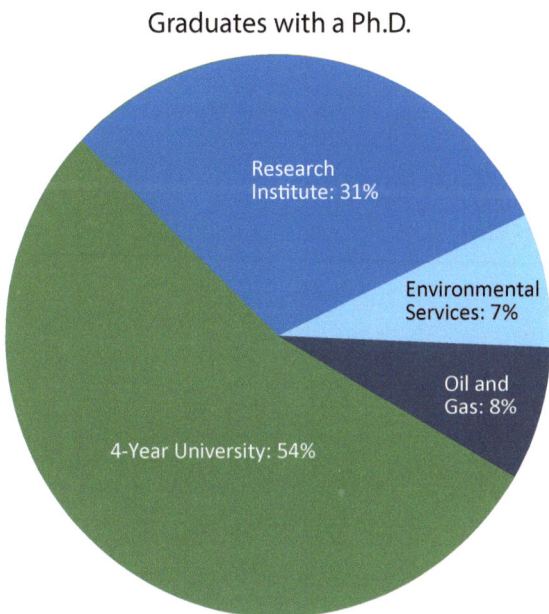

Graduates with a BA/BS

- Research Institute: 7%
- Information Services: 4%
- Mining: 5%
- NonProfit: 5%
- Oil and Gas: 5%
- Other Education Services: 5%
- Agriculture/Forestry/Fishing: 2%
- Construction: 2%
- Information Technology Services: 2%
- K-12 Education: 2%
- Utilities: 2%
- Environmental Services: 30%
- Federal Government: 18%
- 4-Year University: 11%

Graduates with a MA/MS

- Environmental Services: 17%
- 2-Year Colleges: 5%
- 4-Year University: 5%
- Construction: 5%
- Mining: 6%
- State/Local Government: 6%
- Federal Government: 28%
- Oil and Gas: 28%

Graduates with a Ph.D.

- Research Institute: 31%
- Environmental Services: 7%
- Oil and Gas: 8%
- 4-Year University: 54%

Graduating students seeking, or have accepted, a position within the geosciences

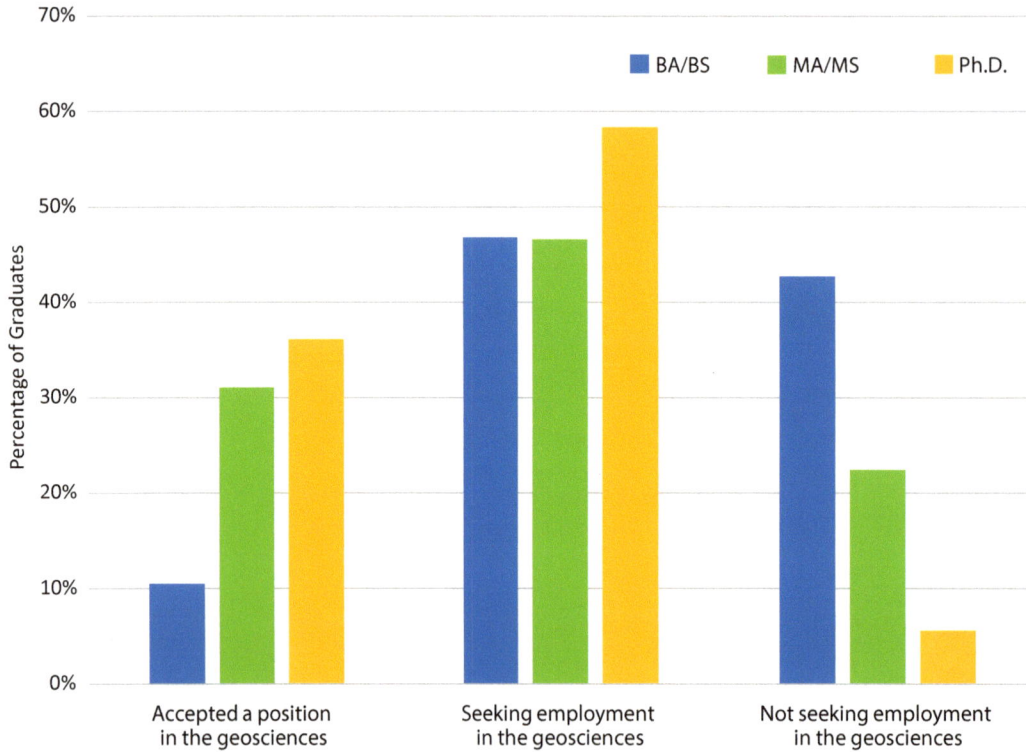

Graduating students seeking, or have accepted, a job within the geosciences by gender

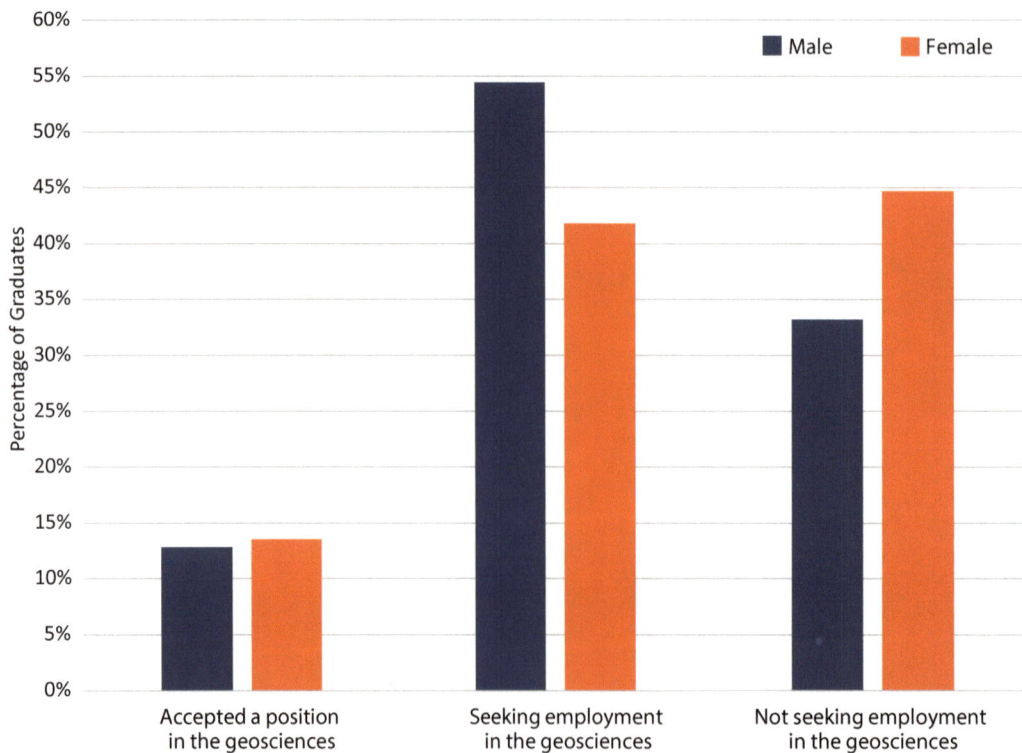

Starting salaries for graduates who accepted a job in the geosciences

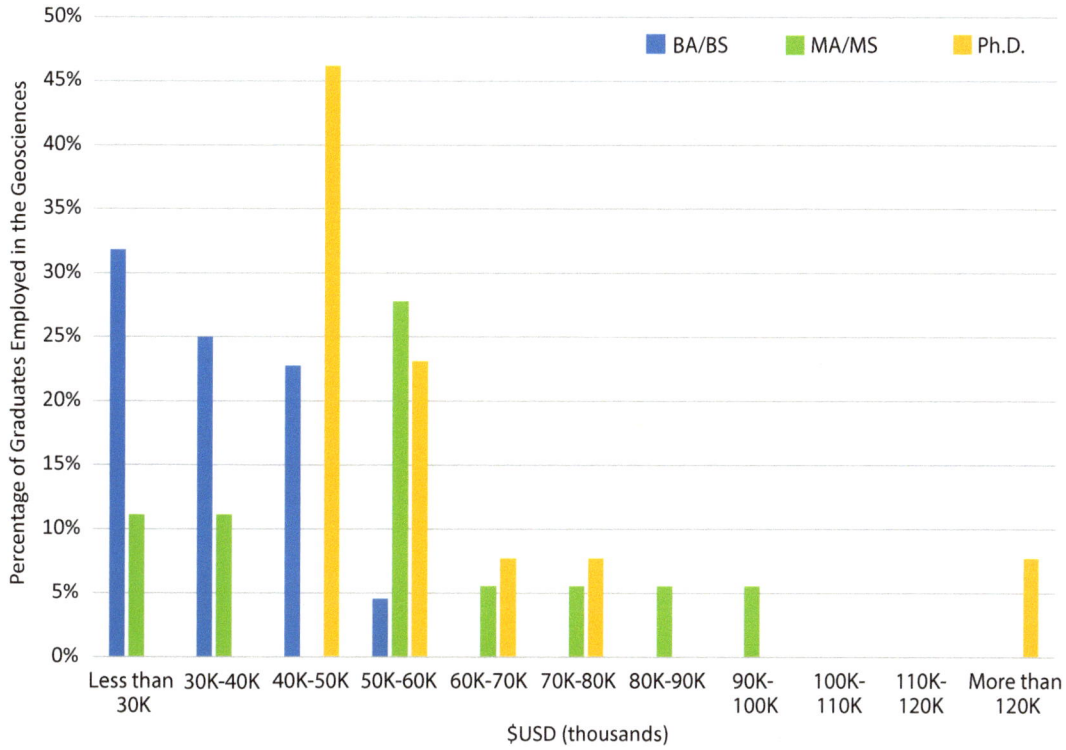

Additional compensation for graduates who accepted a job in the geosciences

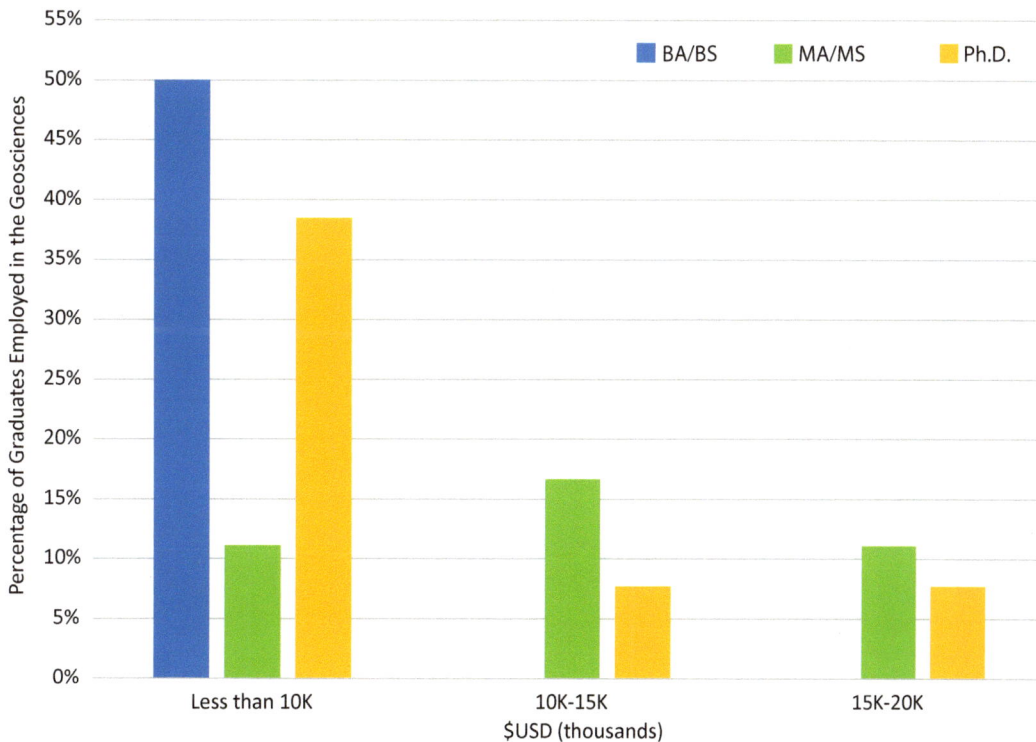

Resources identified by students as useful for finding geoscience jobs

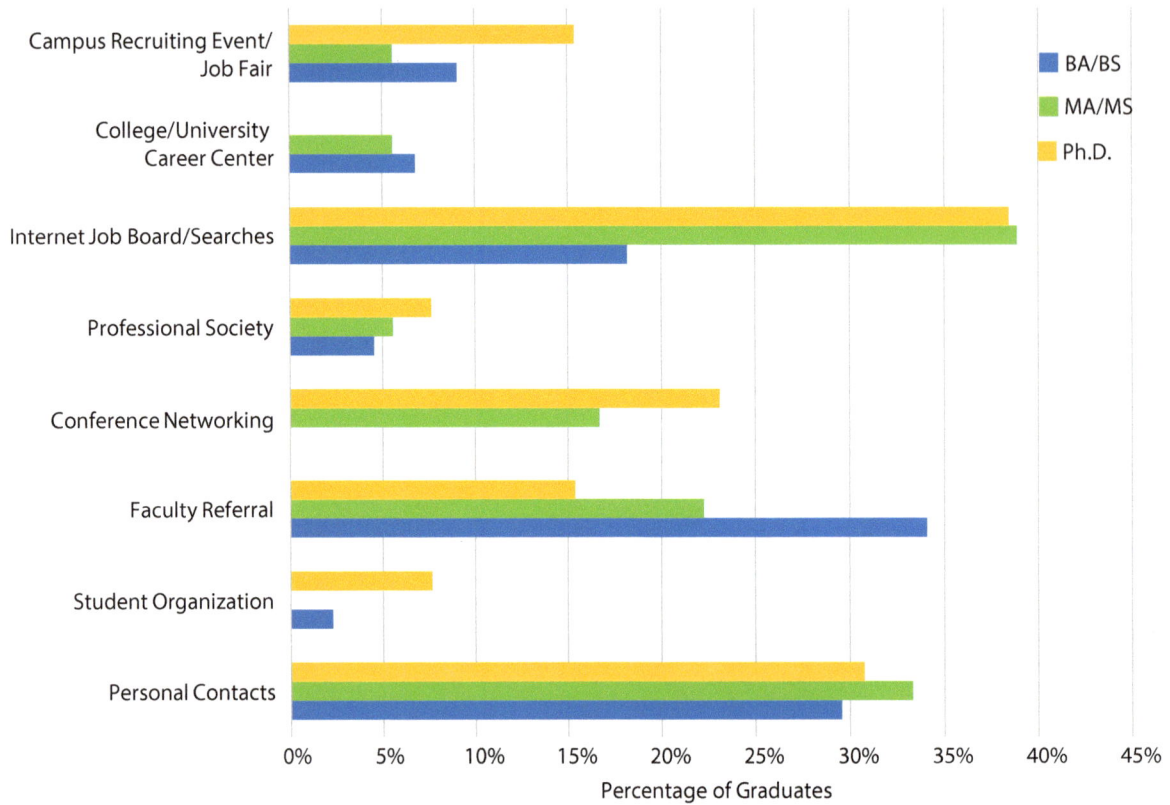

Other job opportunities offered to graduates who accepted a job in the geosciences

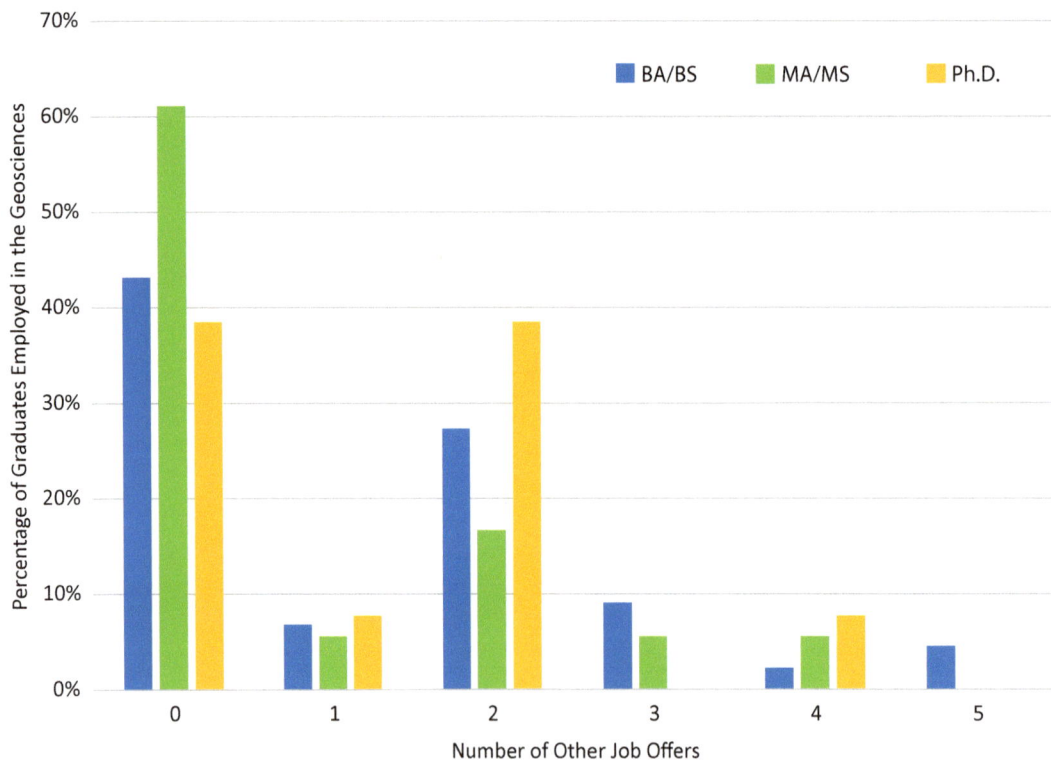

States where graduates found employment in the geosciences

- ☐ 0 Graduates
- ☐ 1-5 Graduates
- ☐ 6-10 Graduates

3 graduates found jobs in international locations

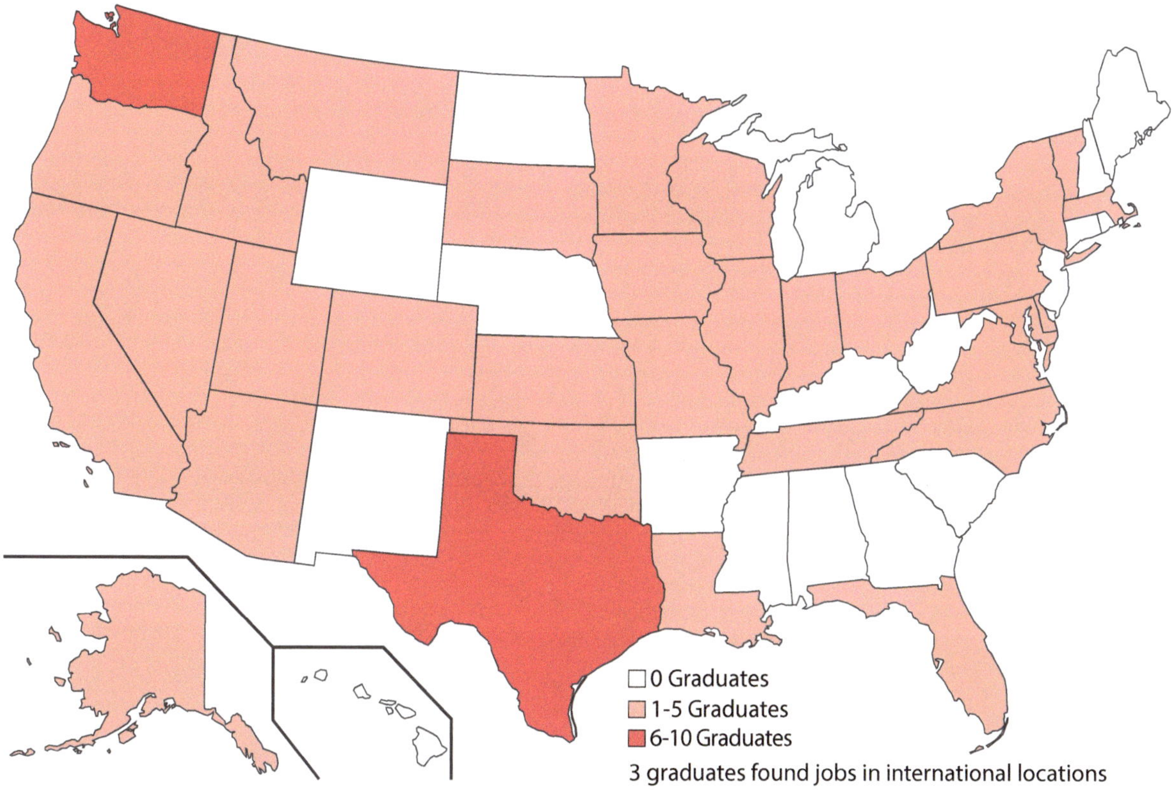

Industries of interest for graduating students seeking a job within the geosciences

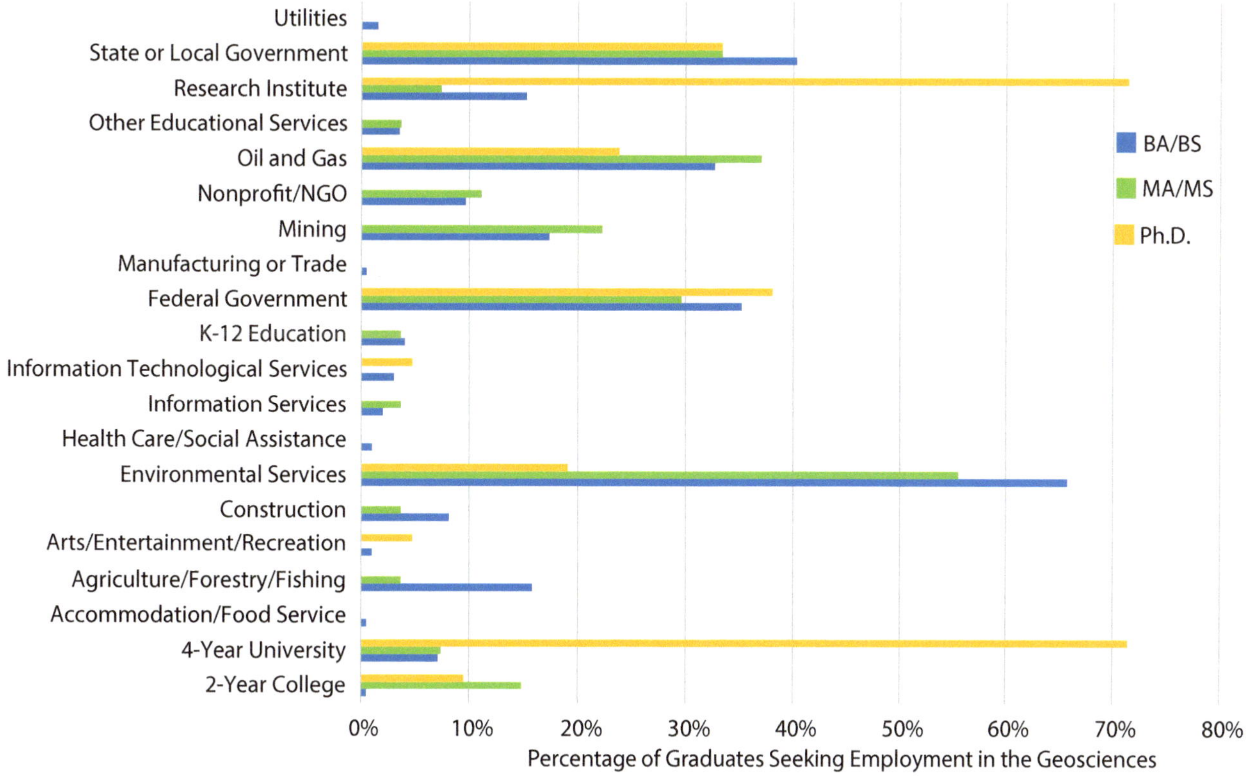

Categories (top to bottom): Utilities, State or Local Government, Research Institute, Other Educational Services, Oil and Gas, Nonprofit/NGO, Mining, Manufacturing or Trade, Federal Government, K-12 Education, Information Technological Services, Information Services, Health Care/Social Assistance, Environmental Services, Construction, Arts/Entertainment/Recreation, Agriculture/Forestry/Fishing, Accommodation/Food Service, 4-Year University, 2-Year College

Legend: ■ BA/BS ■ MA/MS ■ Ph.D.

Percentage of Graduates Seeking Employment in the Geosciences

35

Industries of geoscience graduates' first jobs by degree field for the past five years***

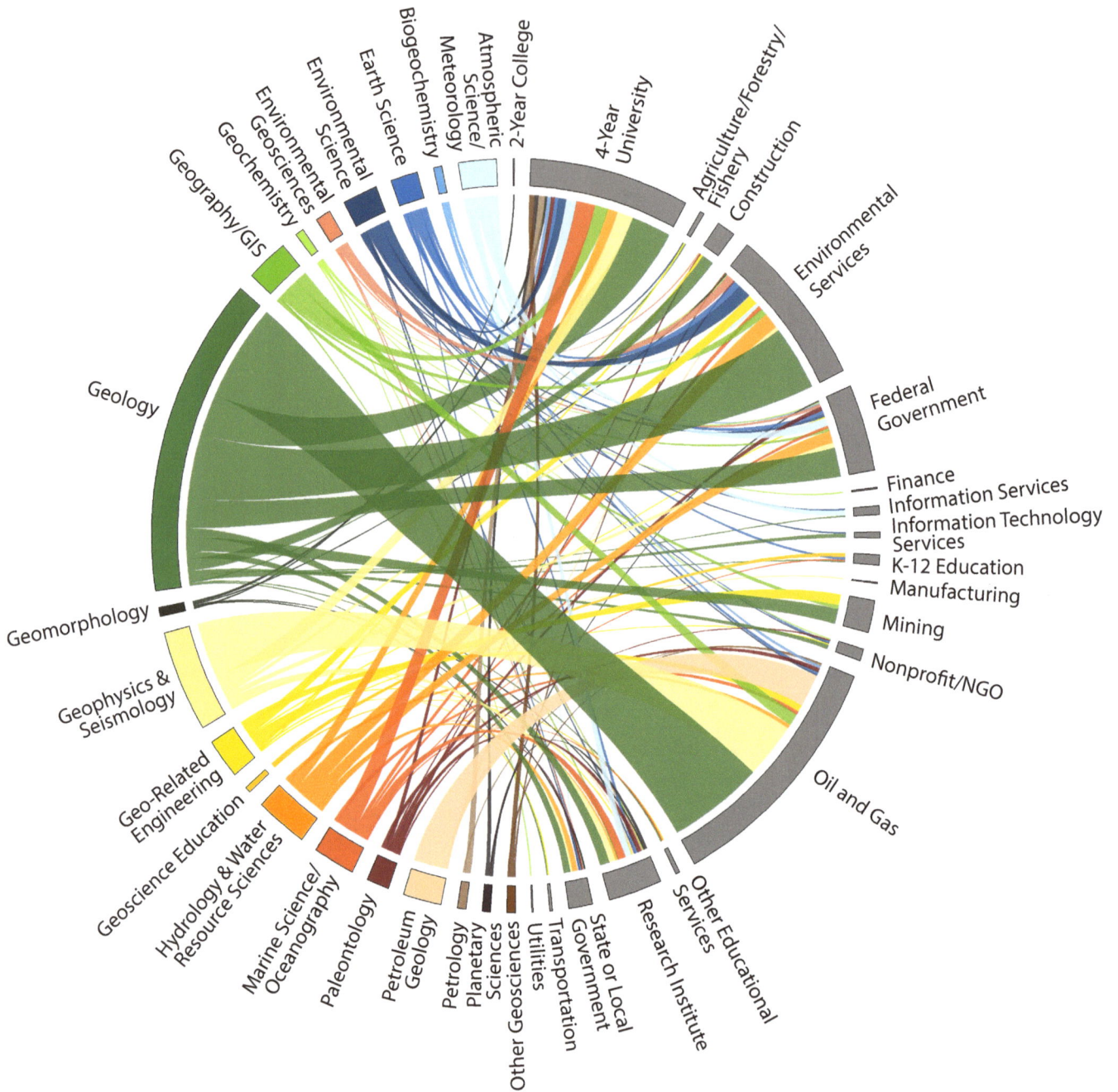

***The code for this visualization was modified from Kyzywinski, M. et al. Circos: an Information Aesthetic for Comparative Genomics.
Genome Res (2009) 19:1693–1645

Victoria Heath for AGI's 2017 Life as a Geoscientist contest

Undergraduate research technicians working with groundwater samples in the Laboratory for Environmental Analysis at Illinois State University.

Future Plans: Working Outside of the Geosciences

Very few recent graduates are seeking or have secured jobs outside of the geosciences. Due to this, the data about these graduates that either accepted or are seeking a job outside of the geosciences show the number of graduates regardless of degree level. In 2017, along with the decrease of graduates finding a job in the geosciences, there was also a decrease in the graduates at all degree levels that found a non-geoscience job at the point of graduation. It is unclear if this hiring was a little slow for the 2017 graduates overall or if we are seeing a hiring trend that will continue with future graduates.

Each year these graduates are asked why they pursued a job outside of the geosciences, and they respond with similar answers. Many graduates seek employment outside of the geosciences because they need a job immediately to help pay bills, were planning to enter the military, wanted to pursue other interests, or take some time before going to graduate school. Often, graduates interested in K-12 education often do not equate teaching earth science as a geoscience job. However, AGI does consider them still within the geosciences community. In 2017, some different responses were provided due to their perceived lack of job options. Multiple graduates felt like they did not have adequate technical or professional skills for the jobs available in the geosciences. Particularly among the bachelor's graduates, some mentioned that all the geoscience jobs they applied for wanted master's level graduates. Multiple bachelor's graduates also mentioned their interest in the geosciences decreased due to experiences endured while working towards their degree. There was also graduates that have a limited definition of geoscience that doesn't cover the various different fields that AGI considers a part of the geosciences. This led to comments expressing their feelings of a limited job pool and to expressing interest in jobs that AGI would consider within the geosciences, even if they do not. Clearly, AGI and the geoscience member societies should help clarify to current students about the options and availability of careers in the geosciences, as well as help them learn how to market themselves for the jobs that they want. Many graduates expressed their frustration with the job hunt, which is understandable, but when AGI surveyed past participants of AGI's Exit Survey earlier this year, they averaged approximately 2 months of time to finding their first job after graduation. Therefore, anyone advising current students applying for jobs should encourage patience and continued effort throughout the job search.

Those graduates that had accepted a job outside of the geosciences were asked to provide more details about their positions. The industries hiring these students included K-12 education, non-profits, arts/entertainment/recreation, finance, military, and accommodation/food service. The majority of these graduates were offered starting annual salaries less than $30,000, and most of these jobs were found through personal contacts, internet searches, and campus recruiting events.

Juliana D'Andrilli for AGI's 2017 Life as a Geoscientist contest

The beginning of molecular-level dissolved organic matter composition analysis: Monitoring the negative mode electrospray ionization of Minnesota peatland porewater dissolved organic matter at the front end of the 9.4 tesla Fourier transform ion cyclotron resonance mass spectrometer at the National High Magnetic Field Laboratory in Tallahassee, Florida.

Graduating students seeking, or have accepted, a job position outside the geosciences

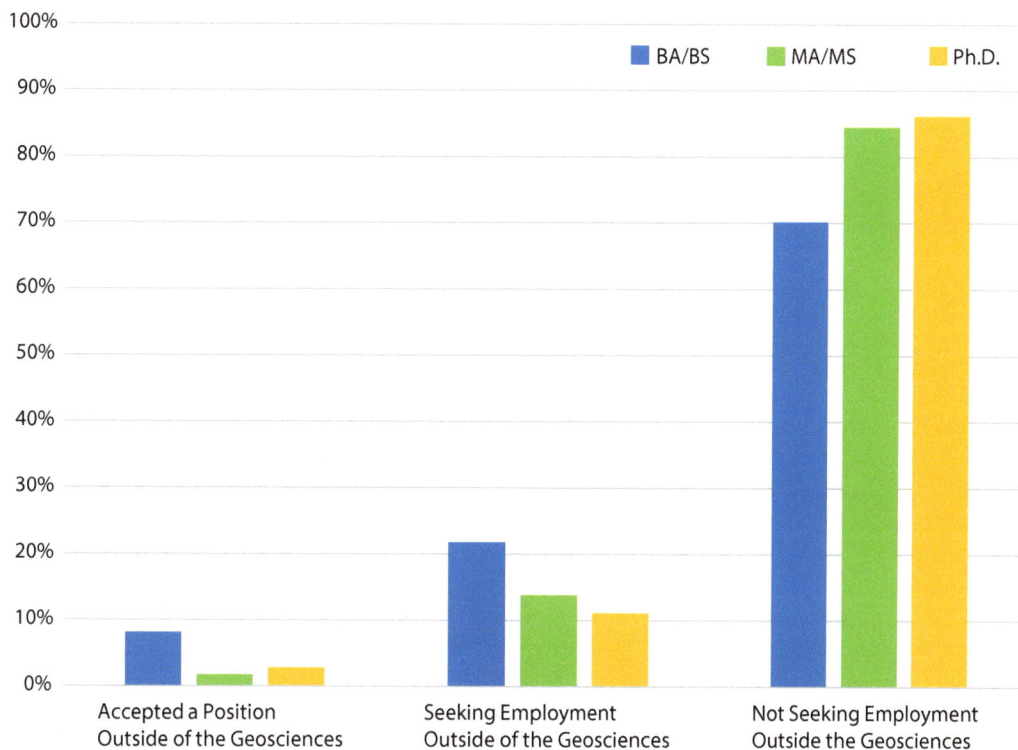

Legend: BA/BS, MA/MS, Ph.D.

Categories: Accepted a Position Outside of the Geosciences; Seeking Employment Outside of the Geosciences; Not Seeking Employment Outside the Geosciences

Industries where graduating students have accepted a job outside the geosciences

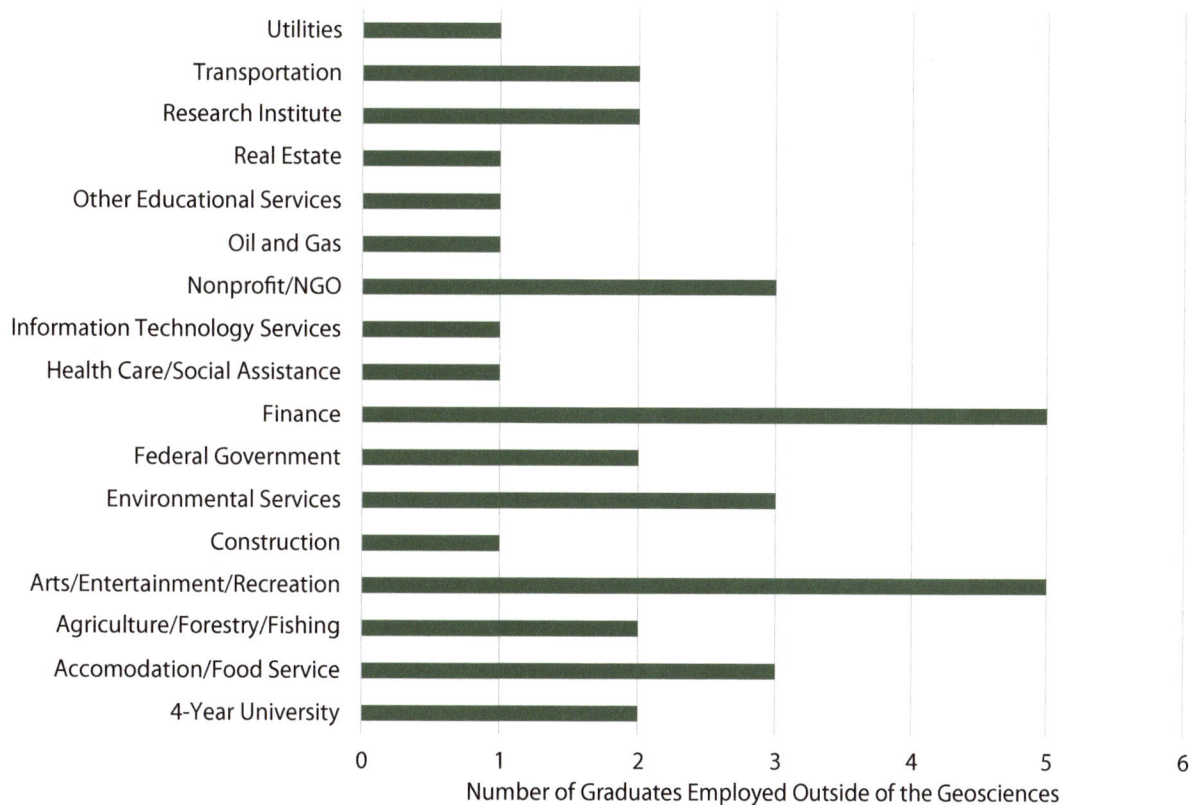

Industry	Number of Graduates Employed Outside of the Geosciences
Utilities	1
Transportation	2
Research Institute	2
Real Estate	1
Other Educational Services	1
Oil and Gas	1
Nonprofit/NGO	3
Information Technology Services	1
Health Care/Social Assistance	1
Finance	5
Federal Government	2
Environmental Services	3
Construction	1
Arts/Entertainment/Recreation	5
Agriculture/Forestry/Fishing	2
Accomodation/Food Service	3
4-Year University	2

Starting salaries for graduating students that accepted a job outside the geosciences

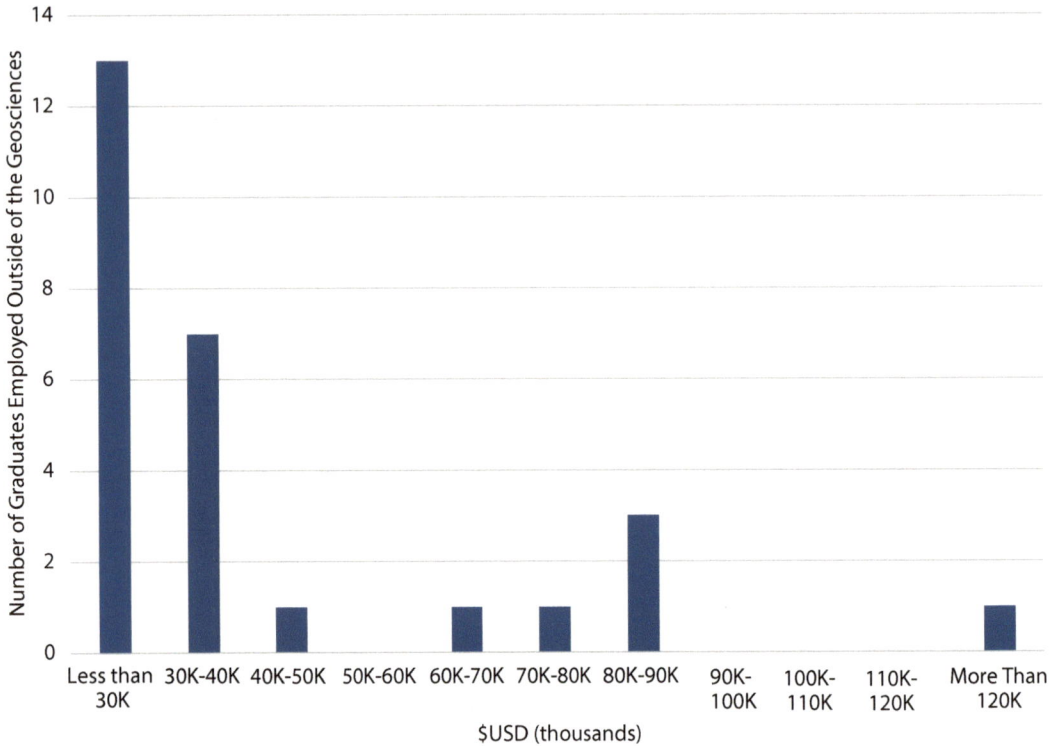

Resources identified by graduating students as useful for finding non-geoscience jobs

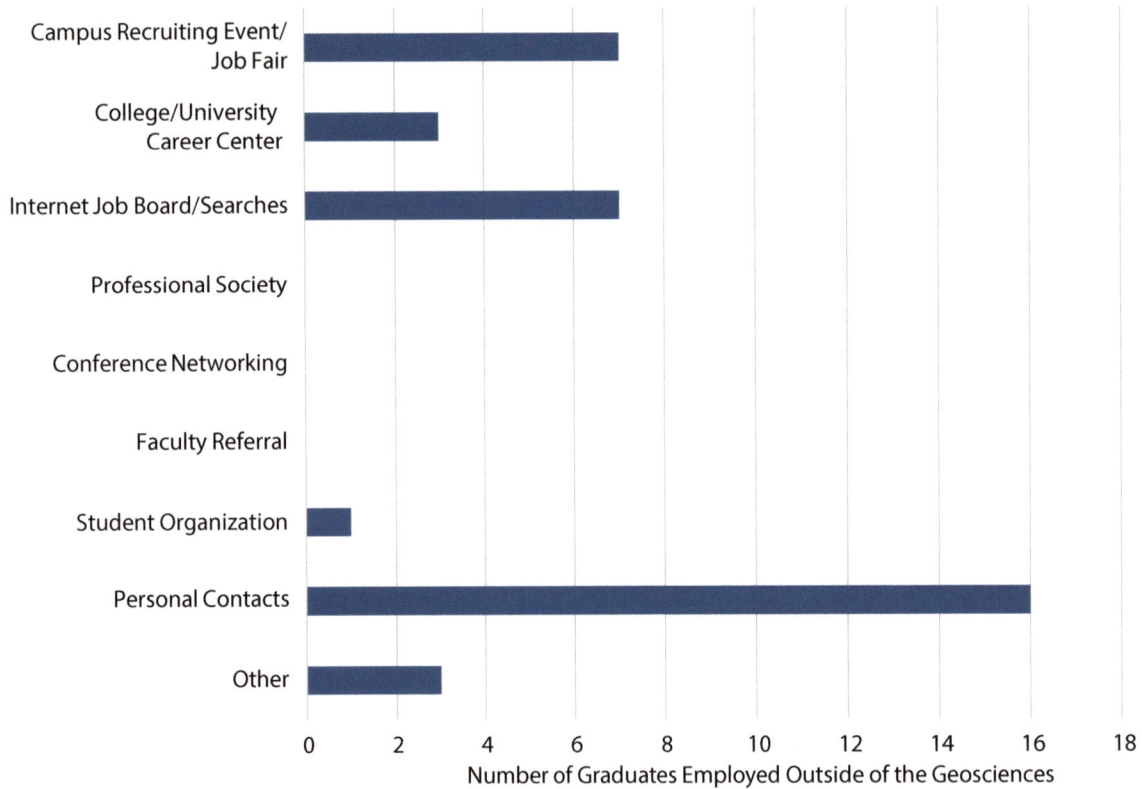

Jackie Buskop for AGI's 2017 Life as a Geoscientist contest

Mazi Onyeali (University of Colorado, Boulder) and Pat Joseph (UWI Seismic Research Center) sampling hydrothermal waters in Watten Waven, Dominica.

Pathway of Preparation for Entering the Geoscience Workforce

The figure below is a Sankey diagram designed to show flow systems visually. In this case, the diagram displays the activities that help develop strong geoscience skills leading to the degree the graduates received and their immediate plans after graduation. It sums up the geoscience experiences of the 2017 graduates also shown through the series of graphs presented earlier in this report to give an overall view of the pathway of preparation for the geoscience workforce among 2017 graduates.

Moving forward, geoscience departments should strive to provide field and research experiences to all geoscience students through their programs, if they do not already, in order for effective development in critical geoscience skills and thinking. Future collaborations between universities, industries, and societies should begin to focus on developing internship-like experiences for more students in order to provide a more realistic understanding of the daily work within the various geoscience industries.

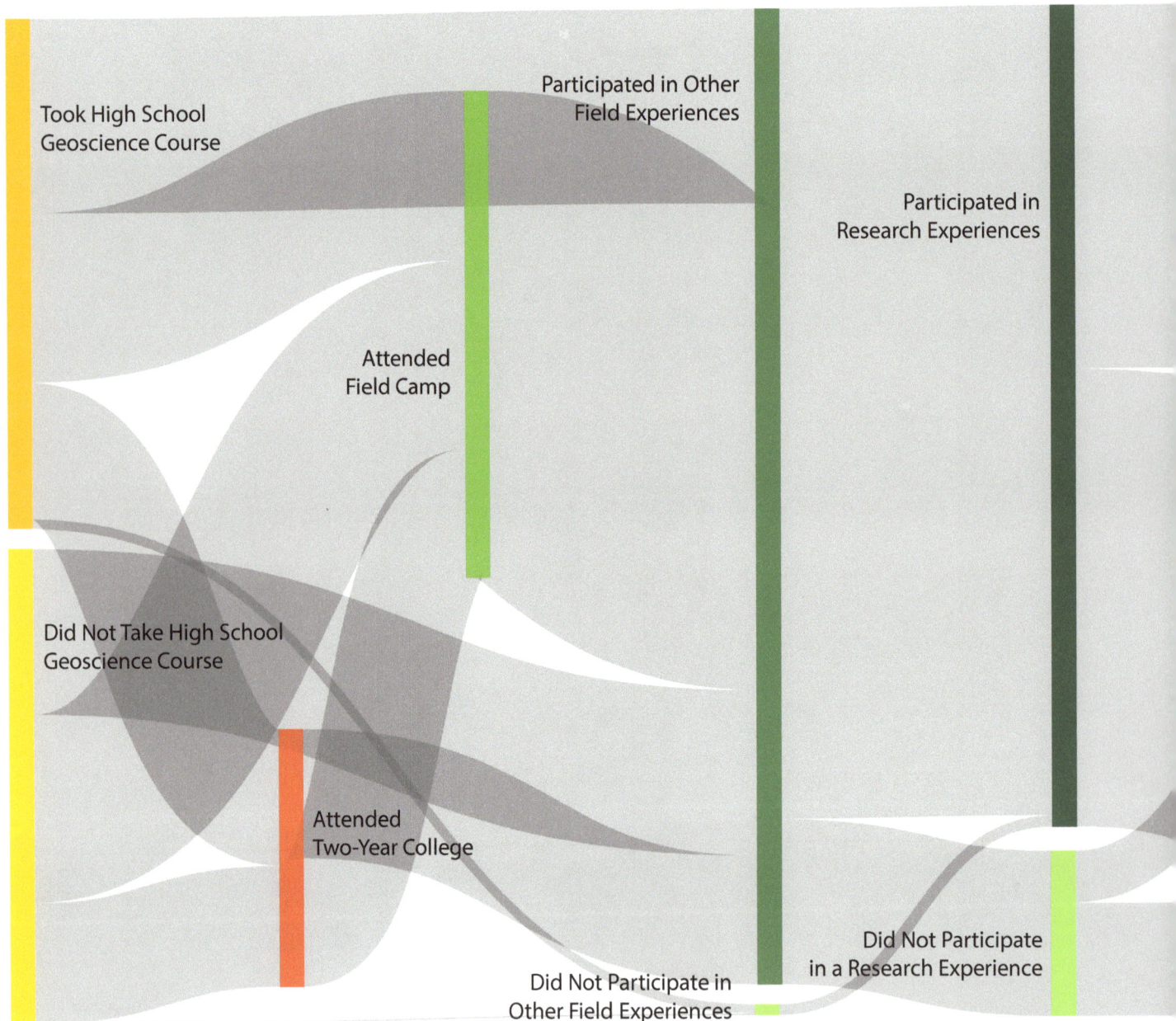

AGI's Geoscience Student Exit Survey will continue to collect data from geoscience graduates each year, and within the year, AGI will reach out to former participants in the survey that are now in the workforce to see how their career pathways have developed. Variations of the survey are currently given in Canada and the UK with plans to expand to other countries in the future.

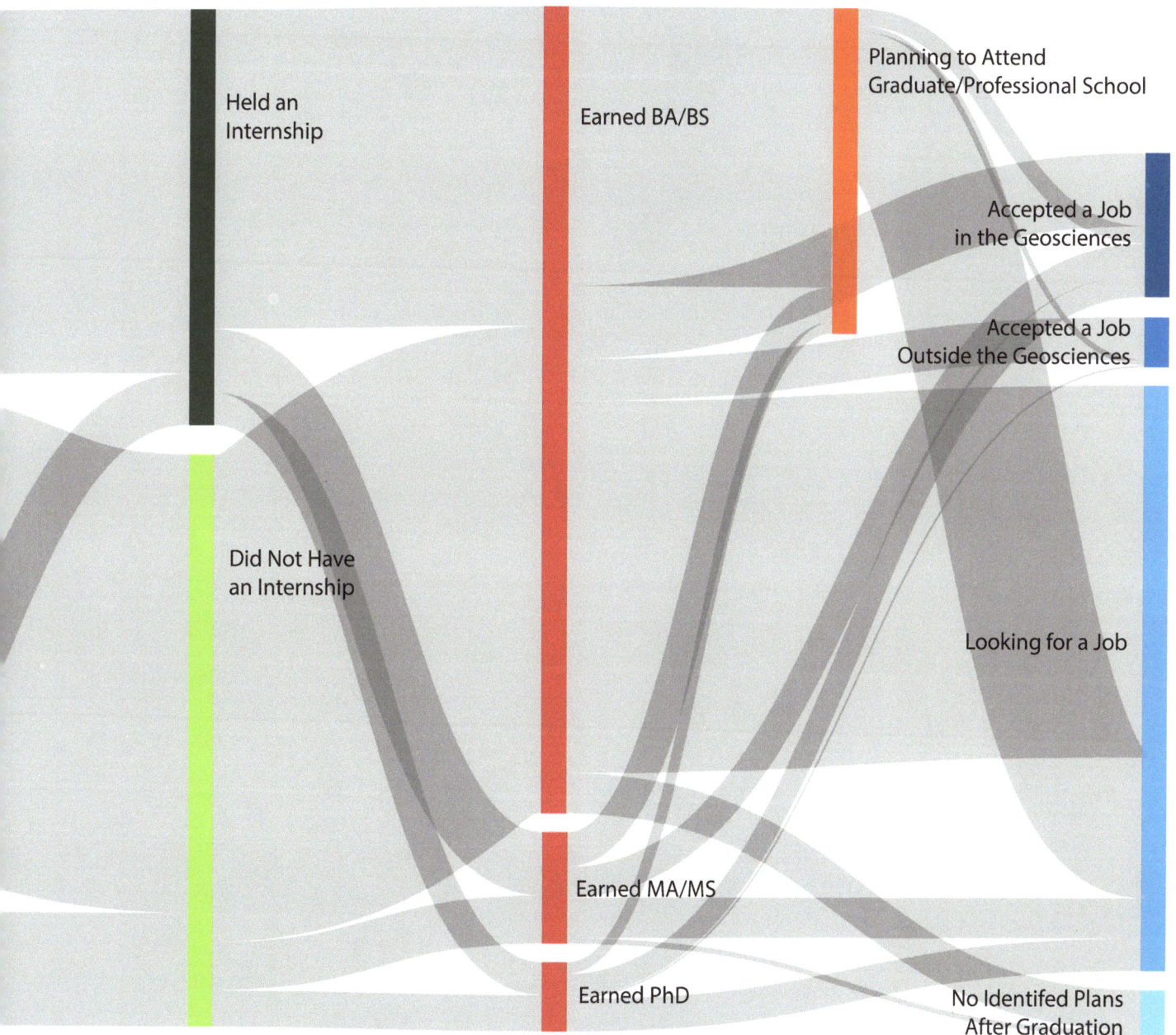

Held an Internship

Did Not Have an Internship

Earned BA/BS

Earned MA/MS

Earned PhD

Planning to Attend Graduate/Professional School

Accepted a Job in the Geosciences

Accepted a Job Outside the Geosciences

Looking for a Job

No Identifed Plans After Graduation

Appendices

Distribution of participating graduating students and departments*

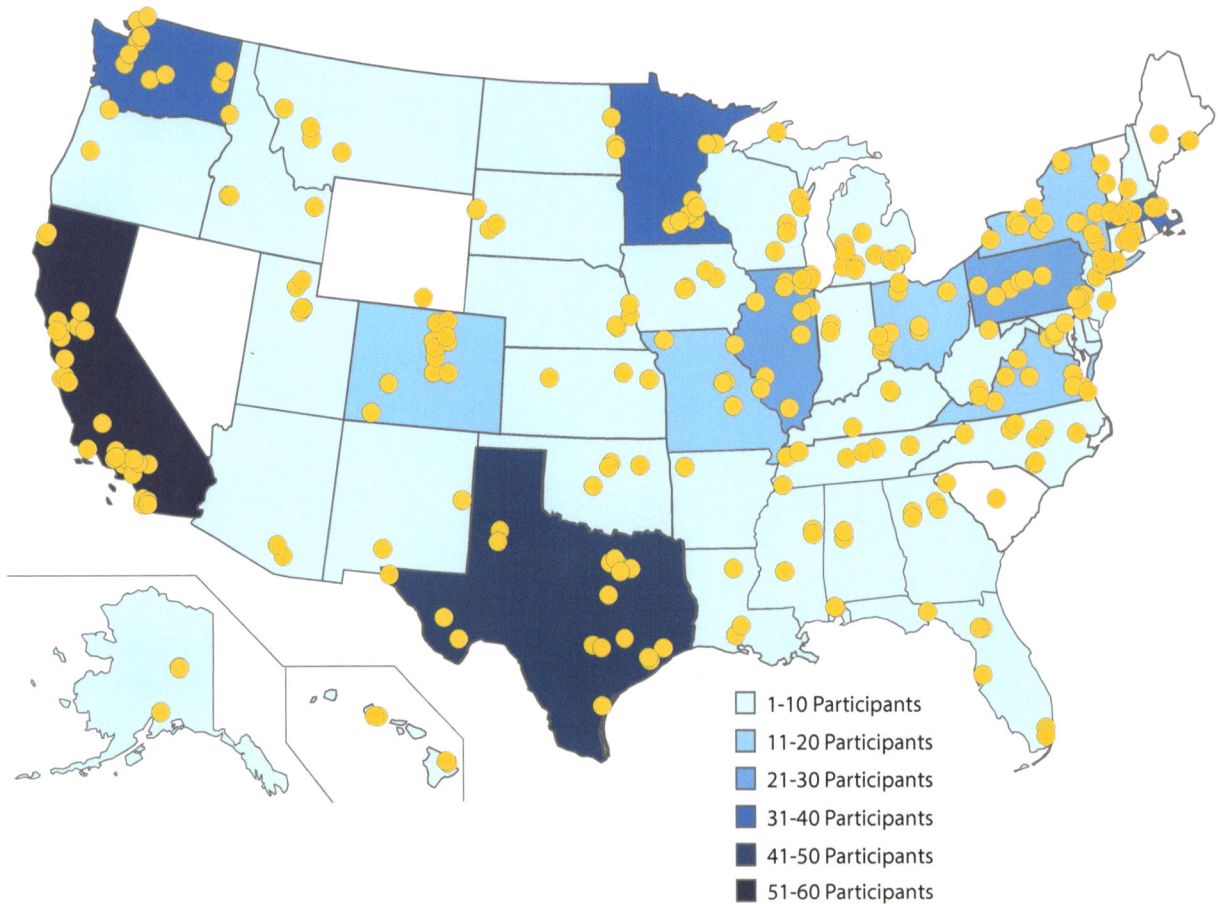

- ☐ 1-10 Participants
- ☐ 11-20 Participants
- ☐ 21-30 Participants
- ☐ 31-40 Participants
- ☐ 41-50 Participants
- ☐ 51-60 Participants

Appendix I

The following is a list of all the institutions and departments with graduating students that took AGI's Geoscience Exit Survey in the 2016-2017 academic year.

University, Department

Amherst College, Department of Geology

Appalachian State University, Department of Geology

Augustana College, Department of Geology

Barnard College, Department of Environmental Sciences

Black Hills State University, Department of Environmental Physical Sciences

Boise State University, Department of Geosciences

Boston College, Department of Earth and Environmental Sciences

Brigham Young University, Department of Geological Sciences

Brigham Young University-Idaho, Department of Geology

Bryn Mawr College, Department of Geology

California State University-Northridge, Department of Geological Sciences

California State University-San Bernadino, Department of Geological Sciences

Carleton College, Department of Geology

Central Michigan University, Department of Earth and Atmospheric Sciences

Central Washington University, Department of Geological Sciences

College of William and Mary, Department of Geology

Colorado College, Department of Geology

Colorado School of Mines, Department of Geology and Geological Engineering

Colorado State University, Department of Geosciences

Concord University, Department of Physical Sciences

Cornell University, Department of Earth and Atmospheric Sciences

CUNY City College, Department of Earth and Atmospheric Science

Eastern Michigan University, Department of Geography and Geology

Eastern Washington University, Department of Geology

Florida International University, Department of Earth and Environment

Florida State University, Department of Earth, Atmospheric, and Oceanic Sciences

Fort Hays State University, Department of Geosciences

Georgia State University, Department of Geosciences

Grand Valley State University, Department of Geology

Guilford College, Department of Geology and Earth Sciences

Gustavus Adolphus College, Department of Geology

Hope College, Department of Geological and Environmental Sciences

Humboldt State University, Department of Geology

Indiana University of Pennsylvania, Department of Geoscience

Iowa State University, Department of Geological and Atmospheric Sciences

James Madison University, Department of Geology and Environmental Sciences

Kansas State University, Department of Geology

Keene State University, Department of Geology

Kent State University, Department of Geology

Miami University of Ohio, Department of Geology and Environmental Earth Science

Middle Tennessee State University, Department of Geosciences

Millsaps College, Department of Geology

Mississippi State University, Department of Geosciences

Massachusetts Institute of Technology, Department of Earth, Atmospheric, and Planetary Sciences

Montana Tech of the University of Montana, Department of Geophysical Engineering

Montclair State University, Department of Geosciences

New Mexico State University, Department of Geological Sciences

North Carolina Central University, Department of Environmental, Earth, and Geospatial Sciences

North Carolina State University, Department of Marine, Earth, and Atmospheric Sciences

North Dakota State University, Department of Geosciences

Northern Illinois University, Department of Geology and Environmental Geosciences

Northwest Missouri State University, Department of Natural Sciences

Northwestern University, Department of Earth and Planetary Sciences

Ohio State University, School of Earth Sciences

Oklahoma State University, Department of Geology

Old Dominion University, Department of Ocean, Earth, and Atmospheric Sciences

Olivet Nazarene University, Department of Geological Sciences

Pacific Lutheran University, Department of Geoscience

Pennsylvania State University, Department of Geosciences

Pomona College, Department of Geology

Purdue University, Department of Earth and Atmospheric Sciences

Rice University, Department of Earth Science

San Diego State University, Department of Geological Sciences

San Francisco State University, Department of Geosciences

Smith College, Department of Geosciences

Sonoma State University, Department of Geology

South Dakota School of Mines and Technology, Department of Geology and Geological Engineering

Southern Illinois University, Department of Geology

St. Lawrence University, Department of Geology

St. Louis University, Department of Earth and Atmospheric Sciences

Stanford University, Department of Geological Sciences

Stockton University, School of Natural Sciences and Mathematics

Sul Ross University, Department of Biology, Geology, and Physical Sciences

SUNY Fredonia, Department of Geology and Environmental Sciences

SUNY Geneseo, Department of Geological Sciences

SUNY New Paltz, Department of Physics and Astronomy

SUNY Potsdam, Department of Geology

SUNY Oneonta, Department of Earth and Atmospheric Sciences

Temple University, Department of Earth and Environmental Science

Texas A&M University, Department of Oceanography

Texas A&M University-Corpus Christi, Department of Geology

Texas Tech University, Department of Geosciences

The Graduate Center-CUNY, Earth and Environmental Sciences Program

Towson University, Department of Physics, Astronomy, and Geosciences

University of Alabama, Department of Geological Sciences

University of Alaska-Anchorage, Department of Geological Sciences

University of Alaska-Fairbanks, Department of Geosciences

University of Arizona, Department of Hydrology

University of Arkansas, Department of Geosciences

University of California-Berkeley, Department of Earth and Planetary Science

University of California-Davis, Department of Earth and Planetary Sciences

University of California-Los Angeles, Department of Earth, Planetary and Space Sciences

University of California-San Diego, Scripps Institution of Oceanography

University of Cincinnati, Department of Geology

University of Colorado at Boulder, Department of Geological Sciences

University of Connecticut, Center for Integrative Geosciences

University of Dayton, Department of Geology

University of Delaware, Department of Geological Sciences

University of Florida, Department of Geological Sciences

University of Georgia, Department of Geology

University of Hawaii-Hilo, Department of Geology

University of Hawaii-Manoa, School of Ocean & Earth Science & Technology

University of Houston, Department of Earth and Atmospheric Sciences

University of Idaho, Department of Geological Sciences

University of Illinois at Chicago, Department of Earth and Environmental Sciences

University of Illinois, Department of Geology

University of Kansas, Department of Geology

University of Kentucky, College of Agriculture, Food, and Environment

University of Louisiana at Lafayette, Department of Geology

University of Maryland, Department of Geology

University of Massachusetts, Department of Geosciences

University of Miami, Department of Geological Sciences

University of Minnesota, Department of Earth Sciences

University of Minnesota-Duluth, Department of Geosciences

University of Missouri, Department of Soil, Environmental, and Atmospheric Science

University of Montana, Department of Geosciences

University of Nebraska-Lincoln, Department of Earth and Atmospheric Sciences

University of Nebraska-Omaha, Department of Geography/ Geology

University of North Carolina at Pembroke, Department of Geology and Geography

University of Northern Iowa, Department of Earth Sciences

University of Oklahoma, School of Geology and Geophysics

University of Oregon, Department of Earth Sciences

University of South Alabama, Department of Earth Sciences

University of Tennessee, Department of Earth and Planetary Sciences

University of Tennessee at Martin, Department of

Agriculture, Geosciences, and Natural Resources

University of Texas at Arlington, Department of Earth and Environmental Sciences

University of Texas at Austin, Jackson School of Geosciences

University of Texas at Dallas, Department of Geosciences

University of Texas at El Paso, Department of Geological Sciences

University of the Pacific, Department of Earth and Environmental Sciences

University of Toledo, Department of Environmental Sciences

University of Tulsa, Department of Geosciences

University of Washington, Department of Earth and Space Sciences

University of Washington, Department of Oceanography

University of Wisconsin-Green Bay, Department of Geosciences

University of Wisconsin-Oshkosh, Department of Geology

Washington and Lee University, Department of Geology

Wayne State University, Department of Geology

Weber State University, Department of Geosciences

Wesleyan University, Department of Earth and Environmental Sciences

West Virginia University, Department of Geology and Geography

Western Michigan University, Department of Geosciences

Western State Colorado University, Department of Geology

Western Washington University, Department of Geology

Wheaton College, Department of Geology and Environmental Science

Williams College, Department of Geosciences

Appendix II

Carnegie Classifications of Institutions of Higher Learning
(http://carnegieclassifications.iu.edu//resources/links.php)

This classification system was used for some of the analysis of the results of AGI's Geoscience Student Exit Survey. The following are the definitions for the classification system and the participating institutions belonging to each category as defined and categorized by the Carnegie Foundation for the Advancement of Teaching.

Baccalaureate Colleges — Arts & Sciences (Bac/A&S)

Baccalaureate Colleges — Diverse Fields (Bac/Diverse)

Includes institutions where baccalaureate degrees represent at least 50 percent of all degrees but where fewer than 50 master's degrees or 20 doctoral degrees were awarded during the update year. (Some institutions above the master's degree threshold are also included). Excludes Special Focus Institutions and Tribal Colleges.

Institutions in which at least half of the bachelor's degree majors in arts and sciences fields were included in the "Arts and Sciences" group, while the remaining institutions were included in the "Diverse Fields" group. For more information about the identification of baccalaureate colleges, please visit the description of the Basic Classification Methodology (http://carnegieclassifications.iu.edu/methodology/basic.php).

Exit Survey Departments (Bac/A&S):
Amherst College
Barnard College
Bryn Mawr College
Carleton College
Colorado College
Guilford College
Gustavus Adolphus College
Hope College
Millsaps College
Pomona College
Smith College
St. Lawrence University
Washington and Lee University

Wheaton College
Williams College

Exit Survey Departments (Bac/Diverse)*
Augustana College
Brigham Young University-Idaho
Montana Tech of the University of Montana

Master's Colleges and Universities — Larger Programs (M1)
Master's Colleges and Universities — Medium Programs (M2)
Master's Colleges and Universities — Smaller Programs (M3)

Generally includes institutions that awarded at least 50 master's degrees and fewer than 20 doctoral degrees during the update year (with occasional exceptions). Excludes Special Focus Institutions and Tribal Colleges.

For more information about the determination of the size of master's programs, please visit the description of the Basic Classification Methodology (http://carnegieclassifications.iu.edu/methodology/basic.php).

Exit Survey Departments (Master's/L):
Appalachian State University
California State University-Northridge
California State University-San Bernadino
Central Washington University
CUNY City College
Eastern Washington University
Fort Hays State University
Grand Valley State University
James Madison University
North Carolina Central University
Northwest Missouri State University
Olivet Nazarene University
Sonoma State University
Stockton University
Sul Ross University
SUNY New Paltz
Towson University
University of Alaska-Anchorage
University of Minnesota-Duluth
University of North Carolina at Pembroke
University of Northern Iowa

University of Wisconsin-Oshkosh
Weber State University
Western Washington University

Exit Survey Departments (Master's/M):

Eastern New Mexico University
Humboldt State University
Pacific Lutheran University
SUNY Fredonia
SUNY Potsdam
University of Tennessee at Martin

Exit Survey Departments (Master's/S):

Black Hills State University
Concord University
Keene State College
SUNY Geneseo
SUNY Oneonta
University of Hawaii-Hilo
University of Wisconsin-Green Bay
Western State Colorado University

Doctoral Universities-Highest Research Activity (DU/R1)

Doctoral Universities-Higher Research Activity (DU/R2)

Doctoral Universities-Moderate Research Activity (DU/R3)

Includes institutions that awarded at least 20 research/scholarship doctoral degrees during the update year (this does not include professional practice doctoral-level degrees, such as the JD, MD, PharmD, DPT, etc). Excludes Special Focus Institutions and Tribal Colleges.

Doctorate-granting institutions were assigned to one of three categories based on a measure of research activity. The "Shorthand" labels for the Doctoral Universities were restored in the 2015 update to numeric sequences to denote that each one is based on differences in quantitative levels. For more information about the methodology to determine the level of research activity, please visit the description of the Basic Classification Methodology (http://carnegieclassifications.iu.edu/methodology/basic.php).

Exit Survey Departments (DU/R1):

Boston College
Brigham Young University
Colorado State University
Cornell University
Florida International University
Florida State University
Iowa State University
Kansas State University
Massachusetts Institute of Technology
North Carolina State University
Northwestern University
Ohio State University
Pennsylvania State University
Purdue University
Rice University
Stanford University
Temple University
Texas A&M University
Texas Tech University
The Graduate Center-CUNY
University of Arizona
University of Arkansas
University of California-Berkeley
University of California-Davis
University of California-Los Angeles
University of California-San Diego
University of Cincinnati
University of Colorado at Boulder
University of Connecticut
University of Delaware
University of Florida
University of Georgia
University of Hawaii-Manoa
University of Houston
University of Illinois at Chicago
University of Illinois
University of Kansas
University of Kentucky
University of Maryland
University of Massachusetts
University of Miami
University of Minnesota
University of Missouri
University of Nebraska-Lincoln
University of Oklahoma
University of Oregon
University of Tennessee
University of Texas at Arlington
University of Texas at Austin
University of Texas at Dallas
University of Washington
Wayne State University
West Virginia University

Exit Survey Departments (DU/R2):

Central Michigan University
College of William and Mary
Colorado School of Mines
Kent State University
Miami University of Ohio
Mississippi State University

New Mexico State University
North Dakota State University
Northern Illinois University
Oklahoma State University
Old Dominion University
San Diego State University
Southern Illinois University
St. Louis University
University of Alabama
University of Alaska-Fairbanks
University of Dayton
University of Idaho
University of Louisiana at Lafayette
University of Montana
University of South Alabama
University of Texas at El Paso
University of Toledo
University of Tulsa
Western Michigan University

Montclair State University
San Francisco State University
Texas A&M University-Corpus Christi
University of Nebraska-Omaha
University of the Pacific

Special Focus Institutions — Schools of Engineering (Spec/Engg)

The special-focus designation was based on the concentration of degrees in a single field of set of related fields, at both the undergraduate and graduate levels. Institutions were determined to have a special focus with concentrations of at least 75% of undergraduate and graduate degrees. Excludes Tribal Colleges.

Exit Survey Departments (DU/R3):
Boise State University
Eastern Michigan University
Georgia State University
Indiana University of Pennsylvania
Middle Tennessee State University

Exit Survey Departments (Spec/Engg)*:
South Dakota School of Mines and Technology

*Institutions in this classification were not included in comparisons using the Carnegie Classification system due to the small number of institutions in the Exit Survey belonging to the particular classification.

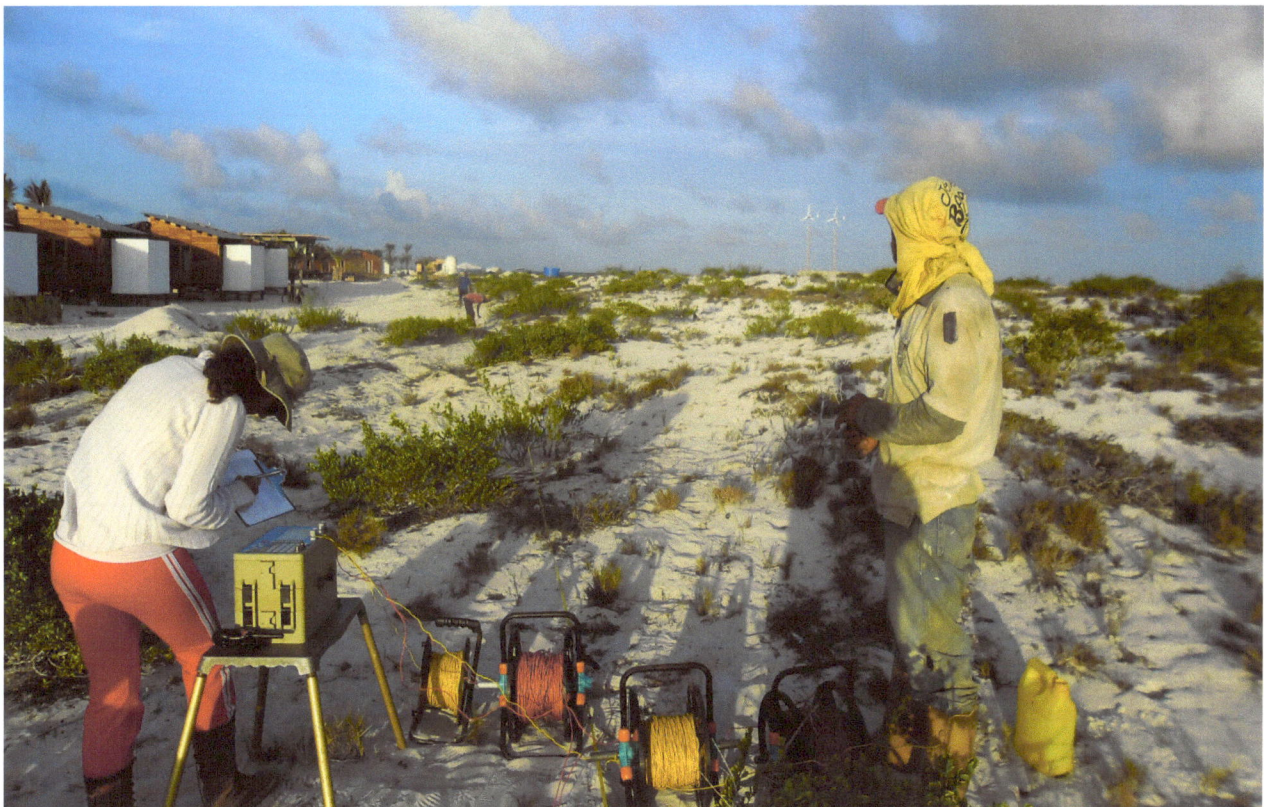

Carianna Herrera for AGI's 2017 Life as a Geoscientist contest
La Tortuga Island, Venezuela. Geophysical acquisition with conventional geoelectric methods for groundwater exploration along the runway of the island.

www.ingramcontent.com/pod-product-compliance
Lightning Source LLC
Chambersburg PA
CBHW052055190326
41519CB00002BA/229